T0190348

Lecture Notes in Computer Science 12390

More information about this series at http://www.springer.com/series/8637

Abdelkader Hameurlain ·
A Min Tjoa · Richard Chbeir ·
Yannis Manolopoulos · Hiroshi Ishikawa ·
Sergio Ilarri · Apostolos Papadopoulos (Eds.)

Transactions on Large-Scale Data- and Knowledge-Centered Systems XLV

Special Issue on Data Management and Knowledge Extraction in Digital Ecosystems

 Springer

Editors-in-Chief
Abdelkader Hameurlain
IRIT, Paul Sabatier University
Toulouse, France

A Min Tjoa
Vienna University of Technology
Vienna, Austria

Guest Editors
Richard Chbeir
University of Pau and Pays de l'Adour
Anglet, France

Yannis Manolopoulos (iD)
Open University of Cyprus
Nicosia, Cyprus

Hiroshi Ishikawa
Tokyo Metropolitan University
Tokyo, Japan

Sergio Ilarri
University of Zaragoza
Zaragoza, Spain

Apostolos Papadopoulos
Aristotle University of Thessaloniki
Thessaloniki, Greece

ISSN 0302-9743   ISSN 1611-3349   (electronic)
Lecture Notes in Computer Science
ISSN 1869-1994   ISSN 2510-4942   (electronic)
Transactions on Large-Scale Data- and Knowledge-Centered Systems
ISBN 978-3-662-62307-7   ISBN 978-3-662-62308-4   (eBook)
https://doi.org/10.1007/978-3-662-62308-4

This Springer imprint is published by the registered company Springer-Verlag GmbH, DE
part of Springer Nature.
The registered company address is: Heidelberger Platz 3, 14197 Berlin, Germany

Preface

In the world of the Internet of Things (IoT), the recent rapid growth and use of connected objects leads to the emergence of virtual environments composed of multiple and independent entities such as individuals, organizations, services, software, and applications sharing one or several missions and focusing on the interactions and interrelationships among them. These digital ecosystems exhibit self-organizing environments where the underlying resources mainly comprehend data management and computational collective intelligence. Due to the multidisciplinary nature of digital ecosystems and their characteristics, it is highly complex to provide a poor understanding as to how managing data will empower digital ecosystems to be innovative and value-creating. The application of Information Technologies has the potential to enable the understanding of how entities request to create benefits and added values, impacting business practices and knowledge. This context introduces many new challenges from different theoretical and practical points of view.

This special issue aims to assess the current status and technologies, as well as to outline the major challenges and future perspectives, related to the data management of digital ecosystems. It includes eight papers that were selected after a very tight peer review, in which each paper was reviewed by three reviewers. Several topics are addressed in this special issue, but mainly: data analysis, information extraction, blockchains, and big data. It is organized as follows.

In the first paper of this special issue, Demetris Trihinas proposes "Interoperable Data Extraction and Analytics Queries over Blockchains." Here, the author explores the explosion of interests by diverse organizations for deploying their services on blockchains to exploit decentralized transaction governance and advanced cryptographic protocols, fostering the emergence of new challenges for distributed ledger technologies (DLTs). The next generation of blockchain services are now extending well beyond cryptocurrencies, accumulating and storing vast amounts of data. Therefore, the need to efficiently extract data over blockchains and subsequently foster data analytics, is more evident than ever. However, despite the wide public interest and the release of several frameworks, efficiently accessing and processing data from blockchains still imposes significant challenges. First, the paper introduces the key limitations faced by organizations in need for efficiently accessing and managing data over DLTs. Afterwards, it introduces Datachain, a lightweight, extensible and interoperable framework deliberately designed to ease the extraction of data hosted on DLTs. Through high-level query abstractions, users connect to underlying blockchains, perform data requests, extract transactions, manage data assets, and derive high-level analytic insights. Most importantly, due to the inherent interoperable nature of Datachain, queries and analytic jobs are reusable and can be executed without alterations on different underlying blockchains. To illustrate the wide applicability of Datachain, we present a realistic use-case on top of Hyperledger and BigchainDB.

The second paper is titled "Exploiting Twitter for Informativeness Classification in Disaster Situations" and authored by David Graf, Werner Retschitzegger, Wieland Schwinger, Birgit Pröll, and Elisabeth Kapsammer. It addresses the problem of disaster management. In essence, this matter urgently requires mechanisms for achieving situation awareness (SA) in a timely manner, allowing authorities to react in an appropriate way to reduce the impact on affected people and infrastructure. In such situations, no matter if they are human induced like shootings or natural ones like earthquakes or floods, social media platforms such as Twitter are frequently used communication channels, making them a highly valuable additional data source for enhancing SA. The challenge is, however, to identify out of the tremendous mass of irrelevant and non-informative social media data which messages are truly "informative", i.e., contributing to SA in a certain disaster situation. Existing approaches on machine learning driven informativeness classification most often focus on specific disaster types, such as shootings or floods, thus lacking general applicability and falling short in classification of new disaster events. Therefore, this paper puts forward the following three contributions: First, in order to better understand the underlying social media data source, an in-depth analysis of existing Twitter data on 26 different disaster events is provided along temporal, spatial, linguistic, and source dimensions. Second, based thereupon, a cross-domain informativeness classier is proposed, focussing not on specific disaster types but rather allowing for classifications across different types. Third, the applicability of this cross-domain classifier is demonstrated, showing its accuracy compared to other disaster type specific approaches.

In the third paper titled "COTILES: Leveraging Content and Structure for Evolutionary Community Detection," Nikolaos Sachpenderis, Georgia Koloniari, and Alexandros Karakasidis address community detection problems. Most related algorithms for online social networks rely solely either on the structure of the network or on its contents. Both extremes ignore valuable information that influences cluster formation. The authors propose COTILES, an evolutionary community detection algorithm, that leverages both structural and content-based criteria to derive densely connected communities with similar contents. Specifically, the authors extend a fast-online structural community detection algorithm by applying additional content-based constraints. They also further explore the effect of structure and content-based criteria on the clustering result by introducing three tunable variations of COTILES that either tighten or relax these criteria. Through an experimental evaluation, they show that the proposed method derives more cohesive communities compared to the original structured one and highlight when the proposed variations should be deployed.

Zahi Al Chami, Chady Abou Jaoude, Bechara Al Bouna, and Richard Chbeir propose "A Weighted Feature-Based Image Quality Assessment Framework in Real-Time" in the fourth paper. Nowadays, social media runs a significant portion of people's daily lives. Millions of people use social media applications to share photos. The massive volume of images shared on social media presents serious challenges and requires large computational infrastructure to ensure successful data processing. However, an image gets distorted somehow during the processing, transmission, sharing, or from a combination of many factors. So, there is a need to guarantee acceptable delivery content, especially for image processing applications. In this paper, the authors present a framework developed to process a large number of images in real time while estimating the image quality. Image quality evaluation is measured based on four methods: Perceptual Coherence Measure, Semantic Coherence Measure, Content-Based Image Retrieval, and Structural Similarity Index. A weighted quality method is then calculated based on the four previous methods while providing a way to optimize the execution latency. Lastly, a set of experiments is conducted to evaluate the proposed approach.

In the fifth paper, "Sharing Knowledge in Digital Ecosystems Using Semantic Multimedia Big Data" is presented by Antonio M. Rinaldi and Cristiano Russo. In this paper, the authors stress the need to use formal representations in the context of multimedia big data due to the intrinsic complexity of this type of data. Furthermore, the relationships between objects should be clearly expressed and formalized to give the right meaning to the correlation of data. For this reason, the design of formal models to represent and manage information is a necessary task to implement intelligent information systems. Approaches based on the semantic web need to improve the data models that are the basis for implementing big data applications. Using these models, data and information visualization becomes an intrinsic and strategic task for the analysis and exploration of multimedia big data. In this paper, the authors propose the use of a semantic approach to formalize the structure of a multimedia big data model. Moreover, the identification of multimodal features to represent concepts and linguistic-semantic properties, relating them in an effective way, will bridge the gap between target semantic classes and low-level multimedia descriptors. The proposed model has been implemented in a NoSQL graph database populated by different knowledge sources. The authors explore a visualization strategy of this large knowledge base and present and discuss a case study for sharing information represented by a model according to a peer-to-peer (P2P) architecture. In this digital ecosystem, agents (e.g. machines, intelligent systems, robots, etc.) act like interconnected peers exchanging and delivering knowledge with each other.

"Facilitating and Managing Machine Learning and Data Analysis Tasks in Big Data Environments Using Web and Microservice Technologies" is proposed as the sixth paper by Shadi Shahoud, Sonja Gunnarsdottir, Hatem Khalloof, Clemens Duepmeier, and Veit Hagenmeyer. Here, the authors address the need for developing easy to use frameworks for instrumenting machine learning effectively for non-data analytics experts as well as novices. Furthermore, building machine learning models in the context of big data environments still represents a great challenge. In this paper, those challenges are addressed by introducing a new generic framework for efficiently facilitating the training, testing, managing, storing, and retrieving of machine learning models in the context of big data. The framework makes use of a powerful big data software stack platforms, web technologies, and a microservice architecture for a fully manageable and highly scalable solution. A highly configurable user interface hiding platform details from the user is introduced giving the user the ability to easily train, test, and manage machine learning models. Moreover, the framework automatically indexes and characterizes models and allows flexible exploration of them in the visual interface. The performance and usability of the new framework is evaluated on state-of-the-art machine learning algorithms: it is shown that executing, storing, and retrieving machine learning models via the framework results in an exceptionally low overhead demonstrating that the framework can provide an efficient approach for facilitating machine learning in big data environments. Configuration options are also evaluated (e.g. caching of RDDs in Apache Spark) based on their affect runtime performance. Furthermore, the evaluation provides indicators for when the utilization of distributed computing (i.e., parallel computation) based on Apache Spark on a cluster outperforms single computer execution of a machine learning model.

The seventh paper is dedicated to "Stable Marriage Matching for Homogenizing Load Distribution in Cloud Data Center," authored by Disha Sangar, Ramesh Upreti, Harek Haugerud, Kyrre Begnum, and Anis Yazidi. Running a sheer virtualized data center with the help of Virtual Machines (VM) is the de facto standard in modern data centers. Live migration offers immense flexibility opportunities as it endows the system administrators with tools to seamlessly move VMs across physical machines. Several studies have shown that the resource utilization within a data center is not homogeneous across the physical servers. Load imbalance situations are observed when a significant portion of servers are either in overloaded or underloaded states. Apart from leading to an inefficient usage of energy by underloaded servers, this might lead to serious QoS degradation issues in the overloaded servers. In this paper, the authors propose a lightweight decentralized solution for homogenizing the load across different machines in a data center by mapping the problem to a Stable Marriage matching problem. The algorithm judiciously chooses pairs of overloaded and underloaded servers for matching and subsequently VM migrations are performed to reduce load imbalance. For the purpose of comparisons, three different greedy matching algorithms are also introduced. In order to verify the feasibility of the provided approach in real-life scenarios, the authors implement the solution on a small testbed. For the large-scale scenarios, they provide simulation results that demonstrate the efficiency of the algorithm and its ability to yield a near-optimal solution compared to other algorithms. The results are promising, given the low computational footprint of the algorithm.

The last paper of this special issue is titled "A Sentiment Analysis Software Framework for the Support of Business Information Architecture in the Tourist Sector" and is written by Javier Murga, Gianpierre Zapata, Heyul Chavez, Carlos Raymundo, Luis Rivera, Francisco Dominguez, Javier Moguerza, and José Maria Alvarez. It addresses a practical problem related to the increased use of digital tools within the Peruvian tourism industry, creating a corresponding increase in revenues. However, both factors have caused increased competition in the sector that in turn puts pressure on small and medium enterprises' (SME) revenues and profitability. This paper aims to apply neural network-based sentiment analysis on social networks to generate a new information search channel that provides a global understanding of user trends and preferences in the tourism sector. A working data-analysis framework is developed and integrated with tools from the cloud to allow a visual assessment of high probability outcomes based on historical data. This helps SMEs estimate the number of tourists arriving and places they want to visit, so that they can generate desirable travel packages in advance, reduce logistics costs, increase sales, and ultimately improve both quality and precision of customer service.

We hope this special issue motivates researchers to take the next step beyond building models to implement, evaluate, compare, and extend proposed approaches.

Many people worked long and hard to help this edition become a reality. We gratefully acknowledge and sincerely thank all the editorial board members and reviewers for their timely and valuable comments and insightful evaluations of the manuscripts that greatly improved the quality of the final versions. Of course, thanks go to all the authors for their contribution and cooperation. Finally, we thank the editors of TLDKS for their support and trust in us, and a special thanks to Gabriela Wagner for her availability and valuable work in the realization of this TLDKS volume.

July 2020

Richard Chbeir
Yannis Manolopoulos
Hiroshi Ishikawa
Sergio Ilarri
Apostolos Papadopoulos

Organization

SI Editorial Board

Sabri Allani	Université de Pau et des Pays de l'Adour, France
Adel Alti	Constantine University, Algeria
Richard Chbeir	Université de Pau et des Pays de l'Adour, France
Joyce El Haddad	UDO, France
Anna Formica	Istituto di Analisi dei Sistemi ed Informatica, CNR, Italy
Anastasios Gounaris	Aristotle University of Thessaloniki, Greece
Michael Granitzer	University of Passau, Germany
Abdelkader Hameurlain	Paul Sabatier University, France
Ramzi Haraty	Lebanese American University, Lebanon
Masaharu Hirota	Okayama University of Science, Japan
Sergio Ilarri	University of Zaragoza, Spain
Hiroshi Ishikawa	Tokyo Metropolitan University, Japan
Lara Kallab	Nobatek, France
Helen Karatza	Aristotle University of Thessaloniki, Greece
Georgia Koloniari	University of Macedonia, Greece
Anne Laurent	University of Montpellier, France
Aristidis Likas	University of Ioannina, Greece
Yannis Manolopoulos	Open University of Cyprus, Cyprus
Apostolos Papadopoulos	Aristotle University of Thessaloniki, Greece
Imad Saleh	University of Paris 8, France
Maria Luisa Sapino	Università degli Studi di Torino, Italy
Joe Tekli	Lebanese American University, Lebanon
Demetris Trihinas	University of Nicosia, Cyprus
Jose R. R. Viqueira	University of Santiago de Compostela, Spain

External Reviewers

Karam Bou Chaaya	Université de Pau et des Pays de l'Adour, France
Elio Mansour	Université de Pau et des Pays de l'Adour, France

Contents

Interoperable Data Extraction
and Analytics Queries over Blockchains

Demetris Trihinas$^{(\boxtimes)}$ (iD)

Department of Computer Science, University of Nicosia, Nicosia, Cyprus
trihinas.d@unic.ac.cy

Abstract. The explosion of interests by diverse organisations for
deploying their services on blockchains to exploit decentralize transaction governance and advanced cryptographic protocols, is fostering the
emergence of new challenges for distributed ledger technologies (DLTs).
The next generation of blockchain services are now extending well beyond
cryptocurrencies, accumulating and storing vast amounts of data. Therefore, the need to efficiently extract data over blockchains and subsequently foster data analytics, is more evident than ever. However, despite
the wide public interest and the release of several frameworks, efficiently
accessing and processing data from blockchains still imposes significant
challenges. This article, first, introduces the key limitations faced by
organisations in need for efficiently accessing and managing data over
DLTs. Afterwards, it introduces Datachain, a lightweight, flexible and
interoperable framework deliberately designed to ease the extraction of
data hosted on DLTs. Through high-level query abstractions, users connect to underlying blockchains, perform data requests, extract transactions, manage data assets and derive high-level analytic insights. Most
importantly, due to the inherent interoperable nature of Datachain,
queries and analytic jobs are reusable and can be executed without alterations on different underlying blockchains. To illustrate the wide applicability of Datachain, we present a realistic use-case on top of Hyperledger
and BigchainDB.

Keywords: Blockchain · Distributed ledgers · Data analytics

1 Introduction

Today, blockchains are penetrating into industry sectors not initially envisioned
when Bitcoin first appeared in 2008 as the technology capable of disrupting
the banking system by introducing virtual cryptocurrencies and decentralized
monetary transactions over distributed ledger technologies (DLTs) [30]. Moving beyond cryptocurrencies, blockchains are now found in healthcare [16], asset
management [7], intelligent transportation services [35], and even disaster relief
systems [10]. Inevitably, both the volume and velocity at which data are being
stored on blockchains is growing at unprecedented rates [9]. Thus, one of the

© Springer-Verlag GmbH Germany, part of Springer Nature 2020
A. Hameurlain et al. (Eds.) TLDKS XLV, LNCS 12390, pp. 1–26, 2020.
https://doi.org/10.1007/978-3-662-62308-4_1

many fields that have discovered a symbiotic relationship with the next generation of blockchain technologies, is (big) data management [29].

With the volume and complexity of blockchain data continuously growing, we are now witnessing research teams exploring the development of analytics capable of harvesting blockchain data for data-driven decision-making, including: privacy and compliance [20], transaction classification [24], user behavior and interactions [11], and pinpointing mischief and criminal activity [6,31]. To accommodate these complex analytics tasks, both blockchain frameworks are now adopting features from distributed data stores and vice-versa [32].

Nonetheless, augmenting blockchains with distributed data stores is not without challenges. Arguably, the greatest difficulty boils down to querying and extracting data, and then passing changes -or new data products- through the "chain" again. To overcome this challenge, current blockchain frameworks are offering low-level programming drivers for users to access data on top of the blockchain fabric. However, this requires knowledge of the complex nature of blockchain internals which not all users have and also requires spending significant amounts of time on infrastructure code. For example, it is the job of a Data Scientist to discover interesting insights from data assets but not working out how to scan, issue, or validate, encrypted transactions. In turn, while the landscape of blockchain frameworks is still fairly open and non-dominant [12], the lack of interoperability between blockchains is a considerable hindrance for users. Specifically, switching from one platform to another requires the re-coding of complex analytic jobs comprising an applications' governance stack [9]. This is why interoperability, along with scalability and sustainability, is an open challenge for blockchains towards which the EU Commission has come out with a report advocating for blockchain standardization [21].

The main contributions of this article are:

- To present the key limitations faced by organisations in need for efficiently accessing and managing data over DLTs in order for analytics insights to be derived in-time.
- To provide a common and interoperable model for accessing and querying data stored over blockchain platforms. The model goes beyond cryptocurrency transactions, supporting data assets in general.
- To introduce Datachain, a lightweight, flexible and open-source[1] framework for the querying and manipulation of data assets over blockchain platforms by abstracting the interaction complexity with blockchains at runtime. Datachain is comprised of: (i) a python library that can be used in native python scripts or even through jupyter notebooks; (ii) a web service for to interact with other services; and (iii) an SQL-like interface. Highlight features of Datachain include: allowing users to perform ad-hoc queries instead of providing a fixed and pre-determined set of queries, asynchronous requests, informative error handling and automatic query response formatting (e.g., json, xml, pandas etc.). Most importantly, queries and analytic jobs are reusable and can be executed without alterations on different underlying blockchains.

---

[1] https://github.com/dtrihinas/datachain.

- To showcase the wide applicability of Datachain, a use-case scenario is introduced over two popular and open-source blockchain platforms that support the storage of data assets (HyperLedger, BigchainDB).

The rest of this article is structured as follows: Sect. 2 presents a background on blockchains and their relationship with distributed databases. Section 3 introduces key challenges for blockchain databases and highlights the problem description. Section 4 introduces our interoperable data asset model, while Sect. 5 introduces the Datachain Framework and presents illustrative queries and examples. Section 6 presents a comprehensive evaluation of our framework, while Sect. 7 introduces related work, before Sect. 8 concludes the paper.

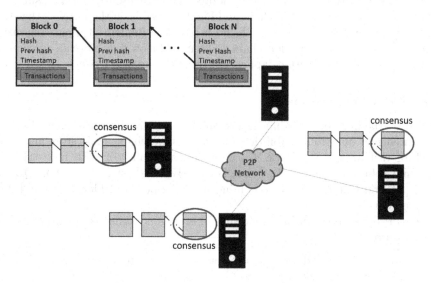

Fig. 1. High-level and abstract overview of a blockchain network

2   Background

2.1   Blockchains

A Blockchain, in the most simplistic form, is a *ledger* that captures timestamped records of transactions with the data hosted in these transactions to be both immutable and encrypted by advanced cryptographic protocols [22]. Figure 1 depicts a high-level overview of a blockchain network, where we observe that a blockchain is essentially comprised of blocks "chained" together to form a linked data structure. A block is comprised of two major parts: (i) the *header*, which captures metadata such as the block hashed signature and a reference to the previous block; and (ii) the *transaction set* encapsulated in the current block. The transaction set of each block forms a Merkle tree to avoid (slow) sequential traversing of lookup queries [27]. Moreover, transactions are considered pseudoanonymous, as all transactions committed to the ledger are publically accessible although involved participants are anonymized.

Although the above linked data structure is sufficient to describe a block-based ledger for timestamped transactions, the true power of a blockchain is harness only when the ledger is *decentralized* alleviating any form of a central authority. Decentralization is achieved by forming a Peer-to-Peer network of computing nodes where each node hosts a replica of the blockchain. Finally, to deal with the synchronization of newly appended transactions to the decentralized ledger, a *consensus mechanism* must be enforced. The underlying algorithm (e.g., proof-of-work, proof-of-stake) adopted in the consensus mechanism may differ from blockchain to blockchain and highly depends on the purpose that the business network is formed [3]. Still, no matter the implementation, consensus defines the set of rules that blockchain nodes must follow to achieve an agreement on the order that transactions appear on the decentralized ledger.

2.2 Blockchain Databases

Fueled by the need of providing secure and decentralized services that also feature intelligent analytic capabilities, inevitably blockchains must store and keep track of vast amounts of data. Towards this, a number of both industry and academic solutions, such as BigchainDB [18], Hyperledger [2], CovenantSQL [8], Storj [33] and Catena [5] are combining both the power of blockchain and distributed databases.

This two-way relationship is fruitful for storage systems as blockchain technologies provide:

- *Secure Storage*, for data assets via advanced cryptographic protocols over decentralized networks;
- *Data immutability*, as validated transactions cannot be tampered with;
- *Transparency*, as all changes to a data asset (e.g., ownership) must be packaged into a transaction;
- *User Accountability*, as changes can be traced back to the point of origin in a possible dispute.

On the other hand, blockchains benefit from what databases do best, *long-term data storage*. An illustrative example is BigchainDB which is marketed as a blockchain database. BigchainDB offers tamper resistance transaction storage for business network entities which is achieved through shared replication across peers, automatic reversion of disallowed updates/deletes and cryptographic signing of all transactions.

3 Motivation and Problem Description

Although integrating blockchains and databases creates a symbiotic relationship to store vast amounts of data, new emerging challenges are appearing which hamper the application in large-scale analytics stacks. First, data extraction from blockchains is significantly slower than databases due to the "chained" data structure, transaction encryption, and the absence of well-defined indexes that can facilitate highly-desired queries [9].

Second, the design and development of query interfaces, along with the efficient mapping for data extraction, are also posing significant challenges for blockchain frameworks. To overcome this challenge, current blockchain frameworks are offering low-level programming drivers for users to access data on top of the blockchain fabric. For instance, BigchainDB features programming drivers in Java and Python to create, submit and extract transactions. To perform more complex queries relevant to the actual stored data assets, one must directly query the underlying database (mongodb) which poses both security and privacy threats. Specifically, the database must be accessible by the issuer of the query and any writes to the database are not passed and tracked through the blockchain. In turn, HyperLedger, now a Linux Foundation project, allows users to access blocks and transactions through the Fabric SDK and data assets through the REST API hosted on top of the blockchain fabric. Unfortunately, more advanced queries are not possible. However, this requires knowledge of the complex nature of blockchain internals which not all users have and also requires spending significant amounts of time on infrastructure code. For example, it is the job of a Data Scientist to discover interesting insights from data assets but not working out how to scan, issue, or validate, encrypted transactions. The importance of the above is also highlighted in the recent Seattle report, where the ACM Fellows identify the design of declarative programming models decoupling the definition of data-oriented tasks from the engineering of the underlying infrastructure as a prominent inhibitor for advancing Data Science[2].

In turn, while the landscape of blockchain frameworks is still fairly open and non-dominant, the lack of interoperability between blockchains is a considerable hindrance for users [12]. Specifically, switching from one platform to another requires the re-coding of complex analytic jobs comprising an applications' governance stack [29]. Thus, any coded analytics jobs would have to be scratched and re-introduced if the organisation is to migrate from one underlying blockchain framework to another [13]. This newly introduced pained, now also mentioned as the analytics stack lock-in, is why interoperability, along with scalability and sustainability, is an open challenge for blockchains. Towards this direction, the EU Commission has recently come out with a report advocating for blockchain standardization [21].

To ease the process of certain analytics tasks, a number of new academia-proposed frameworks have been proposed [4,17,19]. The majority of these frameworks usually compile "snapshots" or "views" of the current state of the business network by periodically polling the underlying blockchain and moving the data to an external database. With the utilization of richer query languages of today's RDBMS solutions, the choice of employing an external database constitutes the querying of the underlying data an easier task. Still, the required data for the

---

[2] https://sigmodrecord.org/2020/02/12/the-seattle-report-on-database-research/.

analytics task is extracted offline and to update the analytics insights, a new snapshot must be re-compiled. This constitutes real-time analysis a slow process which may even occur significant costs for data movement [28]. What is more, there are no guarantees that the data moved to an external database have not been tampered with.

To overcome the aforementioned challenges, we have designed and developed an open and extensible framework satisfying the following objectives:

- **O1:** must integrate a generic and abstract data model extending well beyond cryptocurrencies. The modeling must be alleviated of any blockchain platform specifics (model interoperability).
- **O2:** must support data management operations on top of blockchains (e.g., create/update assets, participants, keys, transactions). This must be supported by hiding the complexity of dealing with blockchain internals (e.g., signing and validating encrypted transactions).
- **O3:** must provide an expressive interface for querying data assets from underlying blockchains. Queries must natively support data filtering, grouping and formatting.
- **O4:** data management operations (O2) and queries (O3) must be issued directly to the blockchain without the requirement of periodically moving data to an external database.
- **O5:** scripted data management operations and queries, must function even when the underlying blockchain changes (analytics interoperability).
- **O6:** must support the integration with other popular Data Science tools (e.g., numpy, pandas, jupyter notebooks, statmodels, etc.).
- **O7:** must provide an adapter interface for integrating new blockchain platforms.

4   Abstract Data Model

In this Section, we introduce a generalized model for abstracting data stored over blockchains.

4.1   Assets

Cryptocurrencies, such as Bitcoin, feature a rather limited data model comprised solely by the value of the cryptocurrency. These cryptocurrencies are created and derive their value directly from the blockchain. Unlike cryptocurrencies, we will consider a much more generalized data model where the smallest reference unit of a data object is an *asset*. An asset can be tangible and physical, such as bicycles, DVDs, houses, cattle, or intangible and virtual, such as stock certificates, insurance contracts or even cryptocurrencies. For physical assets, blockchains are merely a medium to record their existence, evolution and exchanges. Assets are not limited to a single property (e.g., value) but can feature a wide range of properties and be comprised of other assets as well. For example, a bicycle, as depicted below, can have a serial number, manufacturer, color, weight and etc.

```
bike = {
    "identifier" : "bike_serial_number",
    "bike_serial_number": "aXz2r12wQ34",
    "bike_manufacturer": "Raleigh",
    "bike_type": "Road Bike",
    "bike_color": "Red",
    "bike_size": "26",
    "bike_size_units": "in",
    "bike_weight" : 22.7,
    "bike_weight_units": "kg"
}
```

Properties, as the aforementioned, can be *immutable*, meaning they are inchangeable. However, an asset (e.g., bicycle), unlike a cryptocurrency, can have a number of *associated and mutable* properties as well. For example, the owner of a bicycle may be interested in tracking the distance traveled per route using a GPS-enabled monitoring IoT device [28]. Such updates must pass through the blockchain and then linked to the respected bicycle asset for future retrieval. Thus, mutable properties referring to data assets must be both linked to the data asset and afterwards be retrievable when requested. To support this, a blockchain platform could issue a unique identifier to the asset upon creation, albeit users may consider domain-specific identifiers (e.g., a bike's serial number) to more suitable in a business network.

```
bike_associated_data = {
    "bike_serial_number": "aXz2r12wQ34",
    "distance": 5.2,
    "distance_unit": "km",
    "last_upd": "2019-06-14 08:45:38"
}
```

4.2 Participants

Participants are interacting entities of the business service(s) running on top of the underlying blockchain. These participants can own, control, and exercise their rights, privileges, and entitlements over assets. Only the owner of an asset can transfer that asset to another participant. In regards to how current blockchain platforms approach service participants, there are significantly different approaches. For example, in BigchainDB a participant is modelled by a private and public keypair, while in Hyperledger, a participant must own a network authorization card, provided off-the-chain, and can have various (im-)mutable properties just like data assets, as shown below:

```
participant = {
    "userID": "u39492018za1",
    "firstname": "Alice",
    "lastname": "Smith",
```

```
    "joined": "2019-06-01 15:12:04"
}
```

What is more, an asset can be *divisible* for certain blockchain platforms, where an asset is owned by a number of shareholder participants to support partial ownership schemes over divisible assets. Thus, unlike a bicycle which usually has one owner, a football team can be owned by multiple shareholders.

4.3 Transactions

Transactions are what derive value to services run on top of block-chain platforms and are what is recorded on the distributed ledger hosted by the blockchain. Transactions in their most minimal form are comprised of:

- the identifier (e.g., address) of the transaction initiator (e.g, a payment sender);
- the identifier of a transaction consumer (e.g., a payment receiver);
- the commodity (e.g., a bicycle, cryptocurrency, etc.) associated with the transaction.

To support the primary principles of blockchain transactions, we adopt a CRAB model instead of a Create-Retrieve-Update-Delete (CRUD) model which is commonly used by a majority of today's web services. The acronym CRAB refers to Create-Retrieve-Append-Burn transactions over digital assets. Hence, the creation and on-boarding of an asset to the blockchain ledger is associated with a CREATE transaction. Once a CREATE transaction is committed and validated by the underlying blockchain, no changes can be made to the immutable properties of an asset.

A change to an asset can only be associated with the asset's mutable properties, while these changes must be recorded to the ledger. This process is supported via an APPEND transaction. For example, let us assume our bicycle is owned by Alice and that after the completion of a ride, the distance covered must be updated so Alice can keep track of her stats. If this change is not recorded through the ledger then the hashcode of the asset (e.g., Alice's bicycle) will not match due to the change associated with the distance property. Similarly, the transfer of an asset, or shares of an asset, to another participant is also associated with an APPEND transaction, where the transaction consumer is now the new asset owner.

Because of the requirement for every change to be recorded by APPENDing asset associated changes to the ledger, one can keep track of changes and even enforce accountability in cases of dispute. For example, if we consider a shared document, one can RETRIEVE the transactions associated to the asset and always know who made changes, what changes were made and when they were made.

Finally, although in theory an asset cannot be deleted from a blockchain due to the immutability principle, in practice, physical assets can be deleted, discarded or decommissioned. To overcome this issue, one can submit a BURN transaction, which essentially transfers the asset to a vanity address and, thus, essentially nullifying the asset's ownership.

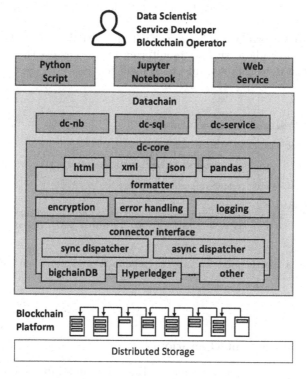

Fig. 2. Datachain high-level architecture

5 Datachain

Figure 2 presents a high-level and abstract overview of the Datachain framework's architecture.

5.1 Users

In general, Datachain targets three user groups which interweave among themselves to achieve their goals:

- **Data Scientists**, who wish to discover hidden insights from the data stored in underlying blockchains. Such users want to quickly perform queries without having to write infrastructure code and ideally work in the environments that they commonly use. Such environments include: python scripts and jupyter notebooks while also utilizing popular analysis libraries such as pandas, numpy, scipy, scikit-learn, etc.

– **Service Developers**, who develop business services for blockchain platforms and wish to ease the interaction with the underlying infrastructure in an interoperable manner. These services may be developed natively through programming adapters or interact with the underlying blockchain through web interface.
– **Blockchain Operators**, who access not the actual data in the blockchain but relevant metadata, such as monitoring logs, to assess the platform performance and detect, faults, misuse or inefficiencies.

```
from Datachain import Datachain

dc = Datachain( 'Bigchaindb',
                endpoints=['http://localhost:9984'],
                params={'api-key': 'AILab-secret-x2019'} )

alice = dc.createParticipant(save=True)

alice = {
  "public_key": "5tHKL2Z6YnwqdMoVSPsHSX5fLGP7uRBLmaUereEJeww6",
  "private_key": "3mPr2MotyJuvyGbr4rUC2bYVPgYe6eKUdsx2LXRQZWsf"
}
```

Fig. 3. Create a participant in jupyter notebook and establish connection to BigchainDB

5.2   Data Management Operations

Figure 3 depicts an example of importing Datachain in a jupyter notebook, connecting to a blockchain platform and creating a new participant. In this example, a connection is established to a locally deployed BigchainDB network, where we also pass a specific API key as an extra security feature for future connections. Afterwards, a new participant to the underlying blockchain is created. Encapsulated under the `createParticipant` method is the interaction with the Datachain `encryption` module which prepares a private/public keypair and then `save`'s the keypair. With the optional `save` parameter, users can securely cache the keypair for subsequent queries in the current session so that the user does not have denote the whereabouts of the keypair for every query. The `encryption` module is also in charge or signing and verifying `CREATE` and `APPEND` transactions, while we note that users do not need to interact with this module, unless they would like to overwrite the current functionality, as all Datachain queries abstract this interaction.

At this point it must be noted that all subsequent operations, either these are asset creations/transfers or queries extracting data from the blockchain, are completely transparent to the underlying blockchain. Thus, if the initial Datachain object were to change and request a connection to a different blockchain (e.g., Hyperledger) all subsequent queries would remain intact requiring no alterations to function (O4). Hence, Fig. 4 shows a Datachain object connecting to

```
dc = Datachain( 'Hyperledger',
                endpoints=['http://localhost:3000'],
                params={ 'api-key': 'AILab-secret-x2019',
                         'remote_tls': True
                       }
              )
```

Fig. 4. Datachain object with connection to hyperledger

Hyperledger while also passing another platform-specific parameter requesting TLS connections for remote queries. For HyperLedger a prominent downside of it's, newly introduced query interface, supported through the REST API, is that it must be programmed prior deployment. This practically means that no ad-hoc and exploratory queries can be supported as the blockchain fabric must be redeployed to acknowledge any new queries not thought at the system design phase. However, this is infeasible for real-time geo-distributed services (e.g., intelligent transportation services) [26]. In contrast, Datachain supports interoperability of the data management and queries through the `connection interface` which abstracts all interactions with underlying blockchains via the data asset model introduced in the previous section. Thus, at any time, ad-hoc queries can be performed and new blockchain providers can be added by simply implementing the data asset model for the respected blockchain. Afterwards, all subsequent high-level operations and queries (e.g., used by `dc-sql`) function without further implementation needs to ease provider on-boarding and Datachain adoption. Hence, all subsequent will be described without mentioning the underlying blockchain platform.

```
bike = { ... }
bike_associated_data = { ... }

#alice credentials already stored, so omitting prev_owner attribute
resp = dc.submitAssetCreateTransaction(bike, ass_data=bike_associated_data)
Datachain>> Asset prepared, signed and verified for submission...
Datachain>> Bigchaindb replied with status: 200, CREATED...
Datachain>> Asset committed to blockchain ledger with transaction id:
9d9fcc8f70b365d2bf8d878312e8fe20ffdda4a9d3dbb0edd6b60c5bbf22e254
```

Fig. 5. Creating a data asset (bicycle) owned by Alice and commit to blockchain using datachain

Figure 5 depicts the creation of a transaction for a newly bought bicycle by Alice. For this transaction, Datachain uses the credentials previously stored when Alice joined the network. We note that the asset properties and associated data have been omitted from the figure and are the same as introduced in Sect. 3. Most importantly, with this **CREATE** operation, Datachain hides the complexity of requiring from users to: (i) prepare the transaction, (ii) cryptographically sign it with the asset owner's (e.g., Alice) private key, (iii) commit the transaction to the blockchain, and (iv) verify the consensus algorithm response.

```
bob = dc.createParticipant()
resp = dc.submitAssetAppendTransaction(bikeid, ass_data=None, prev_trans_id=trans_id,
                                        new_owner=bob, prev_owner=alice)
```

```
Datachain>> Asset prepared, signed and verified for submission...
Datachain>> Bigchaindb replied with status: 200, TRANSFERRED...
Datachain>> Asset committed to blockchain ledger with transaction id:
d88695685142c241d426b0e1bf08dd2575db80dcfbcda70fecafaa0b99ab7574
```

Fig. 6. Transferring a data asset (bicycle) owned by Alice to Bob

In turn, Fig. 6 depicts the transfer of an asset (bicycle) to another owner, Bob. It must be mentioned that although this transaction seems simple, without Datachain it actually requires approximately 102 lines of code and a total of 38 distinct BigchainDB operations. Datachain also provides `error handling` through the respective module. This module parses all errors propagated by the underlying blockchain, prepares informative messages, assigns each error a status code and raises suitable Datachain exceptions. The former are important when embedding Datachain in third-party applications. An example is depicted below where a user (e.g., Alice) attempts to transfer an asset (e.g., her Bicycle) to another participant although the asset has been transferred already (e.g., to Bob). This type of error is commonly referred in blockchain terms as *double spending*.

```
resp = {
    "dc_status_code": "409",
    "dc_msg": "Double spending error... you do not own
        asset with id: aXz2r12wQ34 anymore",
    "trans_id": "5tH...JeAw6",
    "trans_status": "NOT committed"
}
```

```
dc.getAsset(bikeid)
{'bike_serial_number': 'aXz2r12wQ34',
 'bike_manufacturer': 'Raleigh',
 'bike_type': 'Road Bike',
 'bike_color': 'Red',
 'bike_size': '26',
 'bike_size_units': 'in',
 'bike_weight': 22.7,
 'bike_weight_units': 'kg'}
```

Fig. 7. Query for an asset via datachain

5.3   Query Interface

Having introduced some of the data management operations offered by Datachain, we now present some of it's query capabilities. A notable feature of Datachain when performing queries is the `response formatter`. Specifically, when performing a query via Datachain, users can request to receive the result set in various formats. Datachain currently supports `json`, `xml`, `html`, `csv`, python collections (e.g., lists, dicts, tuples) and even NumPy `ndarrays` and Pandas `DataFrames` which are popular tools used by Data Scientists when deriving analytic insights using python and jupyter notebooks. Receiving large volumes of data natively in python, and pandas, can significantly reduce analytics preprocessing time consumed for preparing, shaping and transforming the data. In addition, users can limit the result set size (e.g., `limit=1` means return only the most recent transaction) or receive results sorted, in either `ascending` or `descending` order, based on the transactions' timestamp.

```
tab = dc.getAssetMutableData(bikeid, res_type='pandas')
#distance covered so far
tab['covered_dist'] = tab['distance'].cumsum()
print(tab)
```

	distance	distance_unit	last_upd	covered_dist
0	5.20	km	2019-06-14 08:45:35	5.20
1	4.00	km	2019-06-14 17:24:11	9.20
2	5.20	km	2019-06-15 08:12:10	14.40
3	4.30	km	2019-06-15 16:11:53	18.70
4	13.26	km	2019-06-16 12:18:15	31.96
5	11.20	km	2019-06-17 16:14:50	43.16
6	7.10	km	2019-06-18 18:11:51	50.26
7	4.10	km	2019-06-19 08:11:51	54.36
8	5.80	km	2019-06-19 17:11:51	60.16
9	6.10	km	2019-06-20 08:11:51	66.26
10	9.20	km	2019-06-21 13:11:51	75.46
11	4.80	km	2019-06-22 09:11:51	80.26
12	1.10	km	2019-06-23 15:11:51	81.36
13	3.80	km	2019-06-24 18:11:51	85.16
14	5.30	km	2019-06-25 19:11:51	90.46
15	19.40	km	2019-06-26 14:11:51	109.86

Fig. 8. Query mutable updates on asset (bicycle) and insight extraction (cumsum)

```
▾ #datetime with min distance for alice
  tab.iloc[tab['distance'].idxmin()][['last_upd','distance']]

last_upd    2019-06-23 15:11:51
distance                    1.1
```

Fig. 9. Min distance covered in a single bike route

In what follows, are a number of query examples supported by Datachain. Figure 7 depicts the query of an asset based on the asset's id with the response returned as a python dictionary object as no response type is provided. In turn, Fig. 8 depicts the extraction of updates to an asset's (e.g., Alice's bicycle) mutable properties as a pandas DataFrame and, afterwards, the cumulative sum for the recorded updates is computed and added as a new column to the DataFrame. Another insight that can be derived from the same data is to find the minimum (or maximum) distance covered in a single bicycle route and when it was recorded. This query is depicted in Fig. 9.

Moreover, users can also query for transactions linked to an asset and ownership changes to an asset. The former is depicted in Fig. 10 where the 10 most recent transactions are returned in descending order. The later, Fig. 11 returns the number of ownership changes on an asset along with details of the transaction. For example, Fig. 11 depicts how a Data Scientist would issue a query to obtain the ownership changes to the bicycle initially owned by Alice, sold to Bob and later to Kate. In turn, a feature supported by Datachain but not offered by other programming libraries are native `groupby` queries, where just as with SQL-like DDL, one requests the data grouped by a certain key (or set of keys). Figure 12 depicts such a query where mutable updates to the bicycle asset are first grouped by the date and then daily updates are summed to compute the daily distance covered. Datachain will process this query and the data returned will be in the format requested by the user so that subsequent analysis can be performed without the need for data transformations.

Moreover, to speed up the response time of extracting large volumes of data issued through multiple bulk requests, Datachain offers asynchronous requests. With asynchronous requests, queries for data are submitted in parallel and the client is not blocked awaiting for responses. When results start arriving back to the client, Datachain collects them and provides immediate access while also formatting and filtering responses to the desired type and content. For example, Fig. 13 depicts how easy it is to submit in async mode a large batch of queries requesting for updates to different assets (e.g., bicycles), formatting the results as a pandas DataFrame and then filtering the result set to include only the updates with a distance covered greater than 5 km. Finally, Fig. 14 depicts an example of an SQL-like ad-hoc query performed via the `dc-sql` component to return the 3 most recent updates on an asset's mutable properties. Alas, due to limited space we omit further, and more complex, queries supported by Datachain and refer users to the Datachain examples and tests provided as tests.

```
df = dc.getAssetTransactions(bikeid, res_type='pandas', descending=True, limit=10)
df[['id','operation', 'metadata']]
```

	id	operation	metadata
17	a072298e6b152f89964d9aa8aef233e0e28812bfd44e21...	TRANSFER	None
16	1c3aaece90946cb2216bdef13e9102da1271185de75e77...	TRANSFER	None
15	b080894701e468e1cbca73180d52fe561a0bc216cecbaf...	TRANSFER	{'distance': 19.4, 'distance_unit': 'km', 'las...
14	7d41bf519403f9f1d75869cb6a56eaa128f3e48e966131...	TRANSFER	{'distance': 5.3, 'distance_unit': 'km', 'last...
13	ee5d17eef4f7960677bdff48eccb283fad6032192f45b6...	TRANSFER	{'distance': 3.8, 'distance_unit': 'km', 'last...
12	bdbe35645dab2d5ee7b859ff1a0f29d59719485c793ce8...	TRANSFER	{'distance': 1.1, 'distance_unit': 'km', 'last...
11	e155d39afc5653ff85144f830eb3bff22002387e90c916...	TRANSFER	{'distance': 4.8, 'distance_unit': 'km', 'last...
10	926b8e9cbe851288ca444384f4b8c024c494223e444bcd...	TRANSFER	{'distance': 9.2, 'distance_unit': 'km', 'last...
9	bbbe29b87de590c28dafe1908ad2852b553fdfa8074023...	TRANSFER	{'distance': 6.1, 'distance_unit': 'km', 'last...
8	fc55d1a7d9d675ddbd0fe1aeb51875c3c948dc64338520...	TRANSFER	{'distance': 5.8, 'distance_unit': 'km', 'last...

Fig. 10. Query for 10 most recent transactions in descending order

```
tab3 = dc.getAssetOwnership(bikeid, res_type='pandas')
unique_owners = tab3['current_owners'].map(lambda x: x[0]).unique()
own_num = len(unique_owners)
print('Number of owners for bicycle with id: ', bikeid, ' is ', own_num)
```

```
Number of owners for bicycle with id:   d2d60d80f715f9dbca449e314db09fd9037bba2256ec496f3a9776e111cc9526  is  3
```

Fig. 11. Query for asset ownership changes

```
▾  dc.getAssetMutableData(bikeid, res_type='pandas')
      .groupby('Date')['distance']
      .sum()
```

```
Date
2019-06-14     9.20
2019-06-15     9.50
2019-06-16    13.26
2019-06-17    11.20
2019-06-18     7.10
2019-06-19     9.90
2019-06-20     6.10
2019-06-21     9.20
2019-06-22     4.80
2019-06-23     1.10
2019-06-24     3.80
2019-06-25     5.30
2019-06-26    19.40
Name: distance, dtype: float64
```

Fig. 12. Group by query to aggregate the distance covered per day

```
asset_id_list = [ ... ]
dist_gt_5km = dc.async_query('getAssetMutableProperties',asset_id_list,res_type=pandas)['distance'] > 5.00
```

	distance	last_upd
0	5.20	2019-06-14 08:45:35
2	5.20	2019-06-15 08:12:10
4	13.26	2019-06-16 12:18:15
5	11.20	2019-06-17 16:14:50
6	7.10	2019-06-18 18:11:51
8	5.80	2019-06-19 17:11:51
9	6.10	2019-06-20 08:11:51

Fig. 13. Query for multiple asset updates in async mode

```
dc.sql(SELECT 'distance', 'distance_unit', 'last_upd'
       FROM assets
       WHERE asset_id = 'aXz2r12wQ34'
       ORDER BY DESCENDING
       LIMIT 3
       )
```

```
[{'distance': 5.2, 'distance_unit': 'km', 'last_upd': '2019-06-14 08:45:35'},
 {'distance': 4, 'distance_unit': 'km', 'last_upd': '2019-06-14 17:24:11'},
 {'distance': 5.2, 'distance_unit': 'km', 'last_upd': '2019-06-15 08:12:10'}]
```

Fig. 14. Ad-hoc SQL-like query

6   Evaluation

Datachain is deliberately designed to ease the expressiveness of queries to derive
analytic insights over blockchain stored data. The previous section highlights
the expressiveness and functionality of Datachain. Nonetheless, Datachain can
significantly improve the timeliness of data extraction over blockchains via asynchronous query requests and query operators reducing the result set at the query
level instead of after data retrieval. Thus, in this Section we perform an evaluation of Datachain from a performance perspective.

6.1   Testbed

The testbed of the evaluation, depicted in Fig. 15, is realized by a cluster of 16
servers in an Openstack private cloud. 10 servers are the nodes comprising a
decentralized blockchain network and the other 6 servers act as workload gen-
erators to stress the network and will be simply referred to as client nodes.
Each server is configured with 4VCPU clocked at 2.66 GHz, 4 GB RAM, 260 GB
disk and linked with a 2 Mbps upload and 16 Mbps download network interface.
Between each node we introduce an artificial network latency of ~50 ms. We
opt for these specific capabilities so that the servers resemble actual blockchain
nodes with the network latency giving the testbed a geo-distributed substance
as blockchains nodes are not only decentralized but also scattered across the

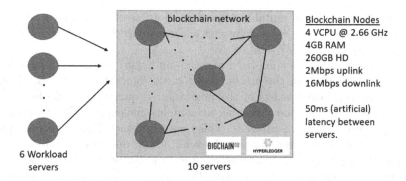

Fig. 15. High-level overview of testbed topology

geographic regions. On each node we deploy and configure BigchainDB[3] and Hyperledger[4] as the blockchain fabric. We note that, when experimenting only one fabric is activated and tested each time. As prerequisites, both platforms require and run over Docker and are built using Docker Compose. The version of BigchainDB is 2.1 and uses MongoDB as the storage backend. The version of Hyperledger is 1.4.1 and uses Composer (v0.20) for the business network modelling and CouchDB as the storage backend.

For the experimentation, we will consider the motivating scenario used throughout the paper. Thus, the blockchain network will function as a decentralized bicycle enthusiast marketplace where users track their assets, stats and can sell, at any time, their bicycles. This, resembles other real-world asset enthusiast marketplaces (e.g., Discogs[5] for vinyl records and CDs). Prior to the evaluation we create a total of 15,000 assets on both blockchains so that the experimentation does not run on an empty data storage backend.

6.2   Experiments

Workloads. To stress the blockchain networks, we develop a blockchain client emulator that accepts a list and percentage ratio of queries to perform. In the experimentation we will consider three distinct workloads:

- **W1.** This workload is characterized as *write-heavy* and is comprised purely of data management operations. Specifically, the workload consists of the following operations: (i) create new asset (10%), (ii) update asset mutable properties (80%), (iii) transfer asset to new owner (8%), and (iv) remove asset from the network (2%).
- **W2.** This workload is characterized as *read-heavy* and is comprised of queries extracting data from the blockchain network. Specifically, the workload consists of the following query types: (i) get asset data by id, (ii) get the 15 most

---

[3] https://www.bigchaindb.com/.
[4] https://www.hyperledger.org/projects/fabric.
[5] https://www.discogs.com.

recent updates to the mutable properties of an asset, (iii) get the 10 oldest transactions for an asset, and (iv) get all ownership changes for an asset. All 4 query types have a 25% ratio.

– **W3.** This workload is comprised of group and aggregation queries to derive analytic insights from the assets stored in the blockchain network. Specifically, the workload consists of the following: (i) query the mutable updates on an asset (bicycle) and subsequently compute the cumulative sum on a certain property (distance covered). This query is depicted in Fig. 8; (ii) query the mutable updates on an asset (bicycle), group updates by date and subsequently aggregate the values to compute the distance covered daily. This query is depicted in Fig. 12; and (iii) This query is similar with the previous, albeit it groups updates by date and bicycle id so that the distance covered daily is derived per bicycle in the case where the user owns more than 1 bicycles. All 4 query types have a 33.3% ratio.

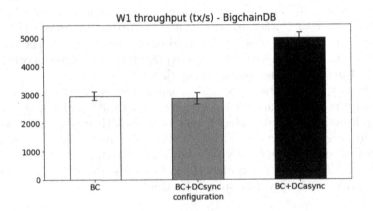

Fig. 16. Throughput W1 - BigchainDB per experiment run

Configurations. In the experimentation we evaluate the following configurations for all workloads. First, we measure the throughput of each blockchain fabric by applying the workloads using the respected fabric's native query environment. For BigchainDB we use it's python driver, while for hyperledger we use it's REST API. Second, we use Datachain in sync mode to apply the respected workload. Third, we use Datachain in async mode to apply the workload. Fourth, we will use LedgerData Refiner (LR), a new framework developed by researchers at Fujitsu Labs, so that Datachain is also compared to another framework [36]. We note that out of the works introduced in Sect. 7, LR is the only solution extending beyond cryptocurrencies and can be used for data assets. Still, LR is bounded to HyperLedger and therefore the experimentation will only be performed for the respective blockchain. What is more, the configurations were run for each workload type 15 times.

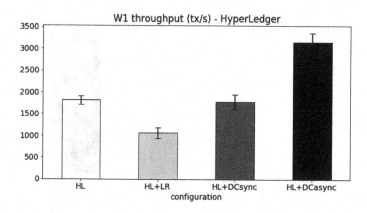

Fig. 17. Throughput W1 - HyperLedger per experiment run

Results. The Figs. 16, 17, 18, 19, 20 and 21 depict the peak performance over a 5 min time interval of the configurations for all three workloads with the experimentation topology comprised of 10 blockchain nodes and 6 workload clients.

From these figures we derive the following insights. First, we observe that in terms of throughput, measured in transactions per second (tx/s), BigchainDB outperforms Hyperledger for both workloads. This is primarily due to the more complex consensus algorithm applied by Hyperledger over all nodes of the blockchain network and the communication overhead as interaction with Hyperledger is performed over HTTP requests. Second, we observe that for W1 and W2, Datachain in sync-mode does not incur a performance penalty (<3%) over using the native programming client for BigchainDB and REST client for Hyperledger. This is a *competitive advantage of Datachain as it not only significantly eases the expression of data management and analytic queries, but it also does this without an additional performance overhead.*

Next, we observe that for all workloads, Datachain (sync mode) performs significantly better than LedgerData Refiner (LR). This is due to Datachain inherent ability to extract and process data in place rather than having to move data to an external data store before queries and data processing can take place. Periodically synchronizing an external data store is a query latency bottleneck. This bottleneck becomes even worse in query operations which must write data back to the blockchain (W1) as the synchronization process is repeated twice. Therefore, Datachain provides over LR a 67% and 56% performance improvement for the W1 (write-heavy) and W2 (read-heavy) workloads, respectively. *By not requiring the periodic synchronization of an external database to store and query extracted blockchain data, Datachain not only can perform real-time queries (no data loss between synchronizations) but it outperforms the current State-of-the-Art which resort to the aforementioned practice.*

What is more, by taking a look at Figs. 20 and 21 (W3), we immediately observe that Datachain (sync mode) provides a performance boost for both blockchain fabrics. The reason for this is inherent to the queries comprising W3

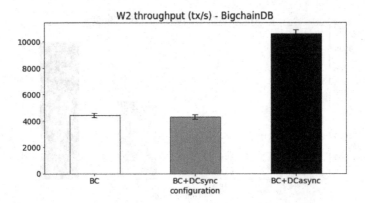

Fig. 18. Throughput W2 - BigchainDB per experiment run

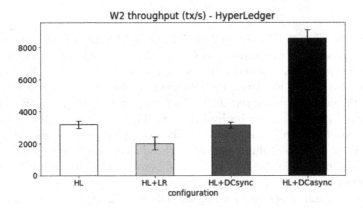

Fig. 19. Throughput W2 - HyperLedger per experiment run

which not only require the extraction of data, but must be formatted, grouped and subsequently aggregation computations are performed. For Datachain this work is inherent to the framework's built-in capabilities but this is not the case for the other under-comparison configurations. Thus, for the blockchain clients significant latency is accumulated for data pre-processing (e.g., for HL extracted data are in JSON format) while LR incurs additional latency due to the external data store synchronization penalty. Specifically, Datachain (sync mode) provides a x1.37 speedup to BigchainDB, a x1.56 speedup to HyperLedger, while the difference between Datachain and LR extends to 286%. *For complex analytics queries requiring data grouping, aggregation, formatting and filtering, Datachain not only eases the expressivity of interoperable queries, but it can provide a performance boost compared to utilizing low-level blockchain clients limited to simple data extraction utilities.*

Finally, we observe that Datachain in async-mode is able to increase the throughput of the underlying networks for both workloads. Specifically, for W1 there is a x1.69 speedup in throughput for BigchainDB and x1.84 for Hyperledger. The maximum speedup per workload client is x4. In turn, for W2 the improvement in throughput is a speedup of x2.4 for BigchainDB and x2.7 for Hyperledger, while for W3 the speedup is x2.79 and x2.85 respectively. As to why Datachain performs better in W2 and W3 than W1, this is primarily due to the fact that async-mode can exploit more the parallelism factor. Specifically, (async) read requests do not wait for a consensus to be agreed between blockchain nodes. Consensus must only be reached when transactions are written to the distributed ledger which is the case for W1. Still, alleviating the client from blocking and awaiting for responses is a non-measurable benefit of exploiting Datachain in write-heavy workloads for blockchains. Thus, *when exploiting async-mode for data extraction to derive analytic insights from blockchain networks, Datachain can boost throughput by providing a speedup of at least x2.4 for read-heavy workloads and at least x1.7 for write-heavy workloads.*

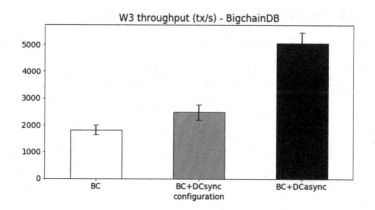

Fig. 20. Throughput W3 - BigchainDB per experiment run

7   Related Work

The following are research and industry-driven frameworks developed to ease -in different ways- the extraction and management of data stored in DLTs. Table 1 summarizes the key features of these frameworks.

Abe [1] is a framework that can be used to periodically crawl the Bitcoin blockchain and store transactions in a relational database for future queries. In turn, McGinn et al. [19], propose a high fidelity graph model and database schema for Neo4J to support offline queries on cryptocurrency transactions, including Bitcoin and Ethereum. On the other hand, Bartoletti et al. [4] introduce a programmable model implemented in Scala to construct offline "views" over cryptocurrency transactions and then perform analytic queries on the extracted data. BlockSci [14] is another notable framework, which provides a

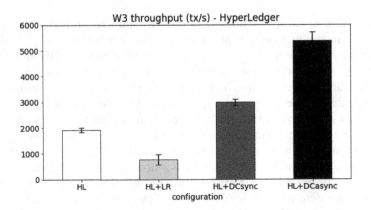

Fig. 21. Throughput W3 - HyperLedger per experiment run

Table 1. Data Management and extraction framework comparison

Framework name	Data management		Data extraction			Data formatting	Apps	Blockchain support
	On chain	Off chain	SQL-like	Prog. Lib	Graphical Env			
Abe		X					Crypto-currencies	BTC
McGinn et al.		X	X				Crypto-currencies	BTC ETH
Bartoletti et al.		X		X (Scala)			Crypto-currencies	BTC ETH
BlockSci		X		X (C++)			Crypto-currencies	BTC
Hawk	X			X (JS)			Smart contracts	ETH
SCILLA	X			X (custom)			Smart contracts	ETH
EtherQL		X	X				Smart contracts	ETH
EQL		X	X			X	Smart contracts	ETH
Splunk fabric		X			X		General	HyperLedger
LedgerData refiner		X			X		General	HyperLedger
Datachain	**X**	**X**	**X**	**X (python)**		**X**	General	**Independent HyperLedger, BigchainDB plugins**

parser to periodically extract transactions from the distributed ledger and then store them in an in-memory database for later analytic queries via a C++ programming library. However, the key limitation of these works is that all require the (periodic) extraction of the data from the underlying DLT so as for processing to be handled "off-chain". In turn, none of these frameworks mention how changes or new insights can be passed through the "chain" again as new data assets.

Moving beyond frameworks supporting offline queries on cryptocurrency transactions, are a number of programming models featuring high-level declarative operators to define complex smart contracts. For instance, Hawk [15] is a programmable framework that applies compiler techniques to hide the complexity of implementing privacy preserving smart contracts. A more generalised approach is introduced by Sergey et al. [23]. Specifically, the authors introduce SCILLA, a high-level and expressive programming model developed for Ethereum to ease the expressiveness in defining smart contracts for digital assets and subsequently verifying initiated transactions. Li et al. [17], introduce EtherQL which is a middleware layer supporting high-level queries for the Ethereum blockchain. However, EtherQL only provides a fixed set of query primitives for analyzing blockchain data, such as range queries and top-k queries. Finally, Splunk [25], the widely known log extraction tool, features a plugin capable of extracting data from HyperLedger business networks. In turn, LedgerData Refiner [36] is a similar (open-source) tool. Both tools only support the graphical exploration of the extracted data and do not support the propagation of new data assets to the blockchain.

Discussion. To date, the research plane for data extraction and management to support the fostering of analytics over blockchains is far from complete. Current frameworks present notable limitations: (i) resort to random scans to discover data; (ii) are offline by periodically scanning the blockchain and then storing off-the-chain the data for subsequent queries; (iii) focus solely on cryptocurrencies, ignoring the industry move to support data assets; and/or (iv) provide a fixed and narrow set of query operators. These limitations can be overcome by utilizing Datachain. Finally, it must be noted that frameworks such as vChain [34] are complimentary to Datachain and together can be used to increase the performance of analytics queries. Specifically, vChain supports verifiable range queries by introducing an accumulator-based authenticated data structure that enables dynamic aggregation over arbitrary query attributes. Incorporating vChain in the Datachain analytics stack can increase the performance of intra- and inter-block aggregation queries. Still, a key limitation of all introduced frameworks, including Datachain, is the absence of support for multi-source queries through the use of join operators. This is understandable when the initial step is first to design a framework to extract and query data, but the next generation of blockchain applications exploring the use of artificial intelligence will require more advanced operators.

8   Conclusion and Future Work

Due to the wide societal interest in blockchain-powered services and the continuous growth of data stored on distributed ledgers, there is a high interest in blockchain analysis leading to an unmet demand for effective query tools. In this article we have presented the challenges which come with querying and managing data assets over blockchains and distributed ledgers in general. To overcome these challenges, we introduce Datachain, a lightweight, flexible and interoperable framework that abstracts the complexity of interacting with blockchains at runtime. Datachain is open-source and provides a toolset comprised of a programming library to query and manage data assets natively in third-party applications, a web service and an SQL-like interface. Key features of Datachain include the ability to define and submit ad-hoc queries instead of being limited to a pre-defined and fixed number of queries, asynchronous requests and informative query handling. Most importantly, queries and analytic jobs are reusable and can be executed without alterations on different underlying blockchains. Although Datachain is designed to ease the management and querying of data assets over blockchains, experiments show that Datachain can also boost throughput when interacting with distributed ledgers and querying for large volumes of data assets and transactional data.

For future work, we intend to integrate Datachain with a dashboard-as-a-service toolset (e.g., Kibana) so as to provide visual exploration of blockchain data by abstracting Datachain's interoperable data exploration and query interface. Moreover, as the current SQL standard is such a powerful object-relational query language, with many query capabilities, we will explore how multi-source queries can be supported in blockchain ecosystems through join operators. In turn, we will research approximate entity matching to improve the query latency over encrypted data transactions through block-based data structures.

References

1. Abe: Block browser for bitcoin. https://github.com/bitcoin-abe/bitcoin-abe
2. Androulaki, E., et al.: Hyperledger fabric: a distributed operating system for permissioned blockchains. In: Proceedings of the Thirteenth EuroSys Conference, EuroSys 2018, pp. 30:1–30:15 (2018)
3. Bano, S., et al.: Consensus in the age of blockchains (2017)
4. Bartoletti, M., Lande, S., Pompianu, L., Bracciali, A.: A general framework for blockchain analytics. In: Proceedings of the 1st Workshop on Scalable and Resilient Infrastructures for Distributed Ledgers, SERIAL 2017, pp. 7:1–7:6 (2017)
5. Catena. https://github.com/pixelspark/catena
6. Chen, W., Zheng, Z., Cui, J., Ngai, E., Zheng, P., Zhou, Y.: Detecting ponzi schemes on ethereum: towards healthier blockchain technology. In: Proceedings of the 2018 World Wide Web Conference, pp. 1409–1418 (2018)
7. Chiu, J., Koeppl, T.V.: Blockchain-based settlement for asset trading. Rev. Finan. Stud. **32**(5), 1716–1753 (2019)
8. CovenantSQL. https://covenantsql.io/
9. Dinh, T.T.A., Liu, R., Zhang, M., Chen, G., Ooi, B.C., Wang, J.: Untangling blockchain: a data processing view of blockchain systems. IEEE Trans. Knowl. Data Eng. **30**(7), 1366–1385 (2018)
10. Castelló Ferrer, E.: The blockchain: a new framework for robotic swarm systems. In: Arai, K., Bhatia, R., Kapoor, S. (eds.) FTC 2018. AISC, vol. 881, pp. 1037–1058. Springer, Cham (2019). https://doi.org/10.1007/978-3-030-02683-7_77
11. Ford, R.A., Swafford, B.L., Shirey, C.B., Moynahan, M.P., Thompson, R.H.: User behavior profile in a blockchain (2018)

12. Gartner: 90% of current enterprise blockchain platform implementations will require replacement by 2021. https://gtnr.it/2OOYwsN

13. Georgiou, Z., Symeonides, M., Trihinas, D., Pallis, G., Dikaiakos, M.D.: StreamSight: a query-driven framework for streaming analytics in edge computing. In: 2018 IEEE/ACM 11th International Conference on Utility and Cloud Computing (UCC), pp. 143–152 (2018)

14. Kalodner, H., Goldfeder, S., Chator, A., Möser, M., Narayanan, A.: BlockSci: design and applications of a blockchain analysis platform. arXiv preprint arXiv:1709.02489 (2017)

15. Kosba, A., Miller, A., Shi, E., Wen, Z., Papamanthou, C.: Hawk: the blockchain model of cryptography and privacy-preserving smart contracts. In: 2016 IEEE Symposium on Security and Privacy (SP), pp. 839–858 (2016)

16. Kuo, T.T., Kim, H.E., Ohno-Machado, L.: Blockchain distributed ledger technologies for biomedical and health care applications. J. Am. Med. Inform. Assoc. 24(6), 1211–1220 (2017)

17. Li, Y., Zheng, K., Yan, Y., Liu, Q., Zhou, X.: EtherQL: a query layer for blockchain system. In: Candan, S., Chen, L., Pedersen, T.B., Chang, L., Hua, W. (eds.) DASFAA 2017. LNCS, vol. 10178, pp. 556–567. Springer, Cham (2017). https://doi. org/10.1007/978-3-319-55699-4_34

18. McConaghy, T., et al.: BigchainDB: a scalable blockchain database (2016)

19. McGinn, D., McIlwraith, D., Guo, Y.: Towards open data blockchain analytics: a bitcoin perspective. Roy. Soc. Open Sci. 5(8), 180298 (2018)

20. Meiklejohn, S., et al.: A fistful of bitcoins: characterizing payments among men with no names. Commun. ACM 59(4), 86–93 (2016). https://doi.org/10.1145/2896384

21. European Union Blockchain Observatory and Forum: Scalability, interoperability and sustainability of blockchains, March 2019

22. Rouhani, S., Deters, R.: Security, performance, and applications of smart contracts: a systematic survey. IEEE Access 7, 50759–50779 (2019)

23. Sergey, I., Kumar, A., Hobor, A.: Scilla: a smart contract intermediate-level language (2018)

24. Spagnuolo, M., Maggi, F., Zanero, S.: BitIodine: extracting intelligence from the bitcoin network. In: Christin, N., Safavi-Naini, R. (eds.) FC 2014. LNCS, vol. 8437, pp. 457–468. Springer, Heidelberg (2014). https://doi.org/10.1007/978-3-662-45472-5_29

25. Splunk: Splunk app for hyperledger fabric. https://splk.it/2uCHG9p

26. Symeonides, M., Trihinas, D., Georgiou, Z., Pallis, G., Dikaiakos, M.: Query-driven descriptive analytics for IoT and edge computing. In: 2019 IEEE International Conference on Cloud Engineering (IC2E), June 2019

27. Szydlo, M.: Merkle tree traversal in log space and time. In: Cachin, C., Camenisch, J.L. (eds.) EUROCRYPT 2004. LNCS, vol. 3027, pp. 541–554. Springer, Heidelberg (2004). https://doi.org/10.1007/978-3-540-24676-3_32

28. Trihinas, D., Pallis, G., Dikaiakos, M.: Low-cost adaptive monitoring techniques for the Internet of Things. IEEE Trans. Serv. Comput. 1 (2018) https://doi.org/ 10.1109/TSC.2018.2808956

29. Trihinas, D.: Datachain: a query framework for blockchains. In: Proceedings of the 11th International Conference on Management of Digital EcoSystems, MEDES 2019, pp. 134–141. Association for Computing Machinery, New York (2019). https://doi.org/10.1145/3297662.3365796

30. Underwood, S.: Blockchain beyond bitcoin. Commun. ACM 59(11), 15–17 (2016)

31. Vasek, M., Moore, T.: There's no free lunch, even using bitcoin: tracking the popularity and profits of virtual currency scams. In: Böhme, R., Okamoto, T. (eds.) FC 2015. LNCS, vol. 8975, pp. 44–61. Springer, Heidelberg (2015). https://doi.org/10.1007/978-3-662-47854-7_4

32. Wang, S., et al.: ForkBase: an efficient storage engine for blockchain and forkable applications. Proc. VLDB Endow. **11**(10), 1137–1150 (2018)

33. Wilkinson, S., Boshevski, T., Brandoff, J., Buterin, V.: Storj a peer-to-peer cloud storage network (2014)

34. Xu, C., Zhang, C., Xu, J.: vChain: enabling verifiable Boolean range queries over blockchain databases. In: Proceedings of the 2019 International Conference on Management of Data, SIGMOD 2019, pp. 141–158. Association for Computing Machinery, New York (2019). https://doi.org/10.1145/3299869.3300083

35. Yuan, Y., Wang, F.Y.: Towards blockchain-based intelligent transportation systems. In: 2016 IEEE 19th International Conference on Intelligent Transportation Systems (ITSC), pp. 2663–2668. IEEE (2016)

36. Zhou, E., Sun, H., Pi, B., Sun, J., Yamashita, K., Nomura, Y.: Ledgerdata refiner: a powerful ledger data query platform for hyperledger fabric. In: 2019 Sixth International Conference on Internet of Things: Systems, Management and Security (IOTSMS), pp. 433–440, October 2019

Exploiting Twitter for Informativeness Classification in Disaster Situations

David Graf[1]([✉]), Werner Retschitzegger[1], Wieland Schwinger[1], Birgit Pröll[2], and Elisabeth Kapsammer[1]

[1] Institute of Telecooperation, Department of Cooperative Information Systems, Johannes Kepler University, Linz, Austria
{david.graf,werner.retschitzegger,wieland.schwinger,
elisabeth.kapsammer}@cis.jku.at
[2] Institute for Application Oriented Knowledge Processing, Johannes Kepler University, Linz, Austria
bproell@faw.jku.at

Abstract. Disaster management urgently requires mechanisms for achieving situation awareness (SA) in a timely manner, allowing authorities to react in an appropriate way to reduce the impact on affected people and infrastructure. In such situations, no matter if they are human-induced like shootings or natural ones like earthquakes or floods, social media such as Twitter are frequently used communication channels, making them a highly valuable additional data source for enhancing SA. The challenge is, however, to identify out of the tremendous mass of irrelevant and non informative social media data those messages being really "informative", i.e., contributing to SA in a certain disaster situation. Existing approaches on machine-learning driven informativeness classification most often focus on specific disaster types, such as shootings or floods, thus lacking general applicability and falling short in classification of new disaster events. Therefore, this article puts forward the following three contributions: First, in order to better understand the underlying social media data source, an in-depth analysis of existing Twitter data on 26 different disaster events is provided along temporal, spatial, linguistic, and source dimensions. Second, based thereupon, a cross-domain informativeness classifier is proposed being not focused on specific disaster types but rather allowing for classifications across different types. Third, the applicability of this cross-domain classifier is demonstrated, showing its accuracy compared to other disaster type specific approaches.

Keywords: Informativeness classification · Disaster related tweets · Cross-domain classification

1 Introduction

Situation Awareness in Disaster Management. In disaster situations, such as *natural disasters* like earthquakes, floods, hurricanes or *human-induced disasters*

© Springer-Verlag GmbH Germany, part of Springer Nature 2020
A. Hameurlain et al. (Eds.) TLDKS XLV, LNCS 12390, pp. 27–55, 2020.
https://doi.org/10.1007/978-3-662-62308-4_2

like shootings or bombings, it is crucial for organizations and authorities to know the extent of the current situation to be able to react in an appropriate way. In many cases, however, detailed information about what happened exactly and what is going on in the affected area is not available [10]. Counteracting this missing *Situation Awareness (SA)*, i.e., "understanding what is happening" [35], in a timely manner is essential to reduce the impact on affected people [4,8,10,29,30].

Social Media for SA. Several studies showed, that social media are a frequently used communication channel even during disaster situations [2,34], thus providing a new source of data for information retrieval tasks [27], more precisely, for gaining SA [4]. In order to exploit the full potential of social media for enhancing SA in disaster management, first of all, social media messages have to be automatically filtered with respect to *informativeness* [28], eliminating *non-related* messages like spam or advertisements and *non informative* ones like emotions or emphatic expressions. Overall, informativeness classification is the crucial basis for all further processing steps, like damage or impact assessment [5]. Especially in scientific research, Twitter[1] or tweets respectively, are a very frequently used data source due to Twitter's APIs, which allow access to real tweet data.

Informativeness of Tweets. The concept of *informativeness* is diverse in its use and discussed in various areas, including, e.g., informativeness of web documents [9], term informativeness [36,37] or informativeness of social media messages such as tweets in areas like news [18] as well as the disaster domain itself [22]. Yet, informativeness is a subjective concept, which heavily depends on the receiver of the information [22]. Since a variety of informativeness definitions exist [7,9,18,19], for the current work we follow the informativeness definition of Olteanu et al. [22] where informativeness of disaster related tweets is captured by "checking whether the tweet contributes to a better understanding of the situation on the ground".

Machine Learning to Classify Tweets. Current approaches for informativeness classification of disaster-related tweets mainly employ *supervised machine learning* [2,12,14,33]. Learning from past events and classification on new events, is, however, quite challenging not least since disaster situations are different in many ways [24]. Thus, most classification approaches focus on informativeness classification for specific types of events, for instance earthquakes or floods, only, lacking general applicability. At the same time, how to get accurate informativeness classification for new disaster events is not yet totally understood due to variations in training data, features, classification algorithms and their settings.

Contributions of Our Work[2]. To address these issues, our contribution is threefold: First, in order to better understand the underlying social media data source

---

[1] http://www.twitter.com.
[2] It has to be noted that a considerably shorter pre-version of this article has already been published in Proceeding of the 10th International Conference on Management of Digital EcoSystems. ACM, Tokyo, Japan, Sept. 2018.

available in disaster situations, a systematic and in-depth analysis of disaster related tweets with respect to informativeness is provided along four different dimensions, covering temporal, spatial, linguistic, and source characteristics. This is done on basis of the CrisisLexT26 [22][3] data set, since this data set provides manually labeled data with respect to informativeness of 26 geographically distributed disaster events of 13 different disaster types. Second, based thereupon, we present a *cross-domain informativeness classifier*, which can be used on new events of various disaster types while being at least as accurate in informativeness classification as disaster type specific ones. Finally, systematic classification experiments are conducted, demonstrating that our classification approach is more accurate than other disaster type specific ones.

Paper Structure. The paper is structured as follows. Section 2 presents the systematic analysis of the CrisisLexT26 dataset being the basis for engineering a cross-domain classifier putted forward in Sect. 3. The applicability of our cross-domain classifier is demonstrated in Sect. 4 on basis of a systematic set of experiments and a comparison of the results with closely related classification approaches. Section 5 reports on related work. Finally, Sect. 6 critically reflects on our approach and discusses future research.

2   Systematic Disaster Data Analysis

Rationale Behind Our Data Analysis. Based on the work of Olteanu et al. [22], the systematic analysis of the CrisisLexT26 presented in the following addresses two main goals: First, we want to analyze similarities and differences of different disaster events and in particular of events of different disasters types. This provides the basis for creating an appropriate training set, which we hypothesize leads to accurate informativeness classification over various disaster types, i.e., addressing the domain adaptation problem [17]. Second, we want to uncover the impact of tweet characteristics with respect to informativeness, to use specifically those having high impact on informativeness as features for classification, which we hypothesize leads to a more accurate informativeness classification compared to other approaches. The systematic analysis is based on 4 different dimensions, comprising i) WHEN a tweet was shared, i.e., *temporal* dimension, ii) WHERE a tweet geographically belongs to, i.e., *spatial* dimension, iii) HOW a tweet is written, i.e., *linguistic* dimension and iv) WHO posts a tweet, i.e., *source* dimension. In order to uncover tweet characteristics correlating with informativeness and their differences with respect to disaster types, analysis of each dimension follows a systematic hierarchical process. To be more specific, the analysis considers, firstly, each dimension on an aggregated level, secondly, on a disaster event level, and, thirdly, in combination with other dimensions. Overall, for each dimension, the analysis is detailed as far as significant differences with respect to informativeness are encountered. In addition to analysis along these dimensions, systematic analysis comprises the investigation of similarity between disaster events as well as their tweets by applying hierarchical clustering.

---

[3] http://www.crisislex.org/data-collections.html.

CrisisLexT26. The analyzed CrisisLexT26 data set contains 28K potentially disaster-related tweets from 26 past disaster events (cf. Table 1), which happened in the years 2012 and 2013. It comprises 13 different disaster types which have been manually labeled by crowd-workers as: "informative and related", "related but not informative", "not related" or "not applicable" and "non informative" [22] which we only considered the former as being informative.

The current section presents, along the four analysis dimensions, our main findings, underpinned with several concrete illustrative examples (cf. Subsects. 2.1–2.4). This is followed by similarity analysis of overall crisis events on the one hand side (cf. Subsect. 2.5) and individual tweets on the other hand side (cf. Subsect. 2.6), before summarizing the data analysis' overall outcomes.

Table 1. CrisisLexT26 dataset [22]

Disaster types	No. types	No. tweets	Category	Development
Bombings	1	1008	Human-induced	Instantaneous
Collapse	1	1255	Human-induced	Instantaneous
Crash	1	1100	Human-induced	Instantaneous
Derailment	3	3050	Human-induced	Instantaneous
Earthquake	4	4464	Natural	Instantaneous
Explosion	2	2111	Human-induced	Instantaneous
Fire	1	1002	Human-induced	Instantaneous
Floods	6	6205	Natural	Progressive
Haze	1	1000	Natural	Progressive
Meteorite	1	1443	Natural	Instantaneous
Shootings	1	1032	Human-induced	Instantaneous
Typhoon	2	2048	Natural	Progressive
Wildfire	2	2400	Natural	Progressive

2.1   Temporal Dimension

By considering the temporal dimension of tweets we want to analyze the evolution of informativeness in time over an entire disaster. For this, it is investigated how far tweet characteristics determined by spatial, source and linguistic dimensions change from the beginning of a disaster to its end. Particular emphasis is put on whether there are differences between disaster events or between disaster types, apart expectable peculiarities induced by the instantaneous or progressive character [22] of certain disaster types like bombings or floods (cf. the "Development" category in Table 1).

Findings. Our findings regarding the temporal dimension can be summarized as follows:

1. Figure 1 visualizes the relation of informative tweets to non informative tweets over time. Overall, this relation stays constant indicated by the dashed gray line. Thus, response time itself, i.e., the period elapsed since the event started and the tweet was sent, is not highly informative for classification.
2. Considering differences in informativeness over time with respect to the *other three analysis dimensions*, only the source dimension shows some peculiarities. While for all events "media", "government" and "NGOs" tend to be much more informative over the entire time period independent of the disaster type, this is not true for sources, "business", "eyewitness" and "outsiders". "Eyewitnesses" over all events, e.g., tend to share, interestingly, in early stages of a disaster more non informative than informative tweets, which turns around after a few days (cf. dashed yellow line in Fig. 1). Since *response time* in combination with other dimensions, like source, shows differences with respect to informativeness, response time is therefore suited to be used as feature for informativeness classification.
3. Analysis shows *no considerable differences* with respect to informativeness between *different disasters types*, except expectable differences with respect to the amount of shared tweets over time between instantaneous disaster types where a majority of tweet communication takes place in the first days and progressive disaster types, where communication is more constant over the entire disaster or correlates with particular occurrences within the event itself, e.g., a rising water level in case of a flood.

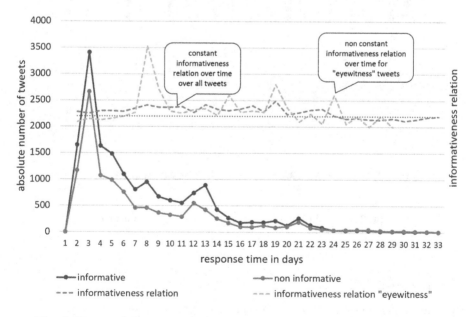

Fig. 1. Temporal dimension - informativeness evolution (Color figure online)

2.2   Spatial Dimension

By considering the spatial dimension of tweets we want to analyze whether there exist differences in informativeness of tweets with respect to the geographic location where the disaster event happened. Due to the geographic distribution of disaster events in the dataset, events were grouped together at a continent level (Asia, Europe, Australia, North- and South-America) based on their country.

Findings. Our main findings towards the spatial dimension can be summarized as follows:

1. The overall relation between informative and non informative tweets in disaster events is *similar* across all continents although there is little variation between single events (cf. Fig. 2). However, the spatial dimension does not provide additional information with respect to informativeness and thus is not used for classification.
2. There are *no considerable differences* between disaster events and between disasters types within one continent, which, as a consequence, support cross-domain classification.

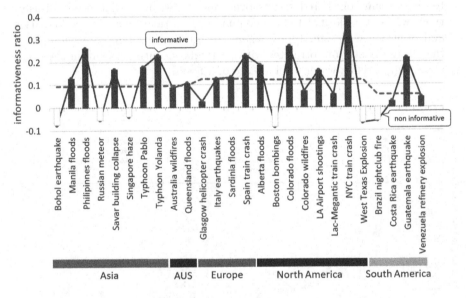

Fig. 2. Spatial dimension - informativeness per continent

2.3   Linguistic Dimension

By considering the linguistic dimension of tweets we want to analyze which linguistic characteristics differentiate informative tweets from non informative

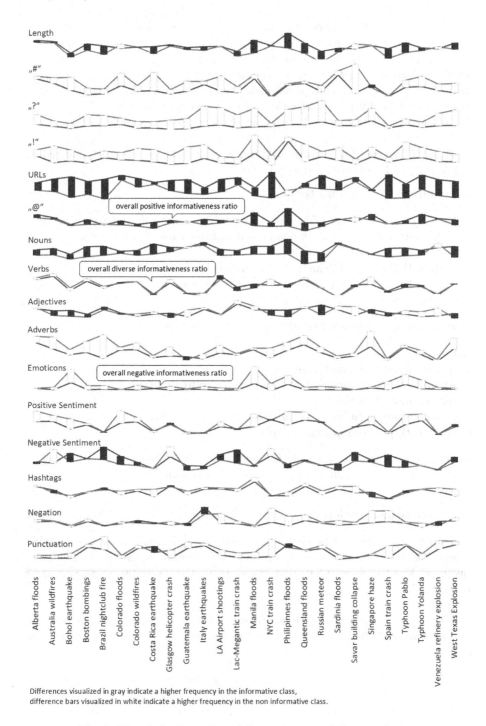

Fig. 3. Linguistic dimension - informativeness of characteristics

ones. Since naturally a variety of linguistic characteristics exist, we focus on those ones which have been already used for classification in other domains such as news [18]. In particular, we analyze i) language, length and sentiment of a tweet, ii) Part-of-Speech (POS) information covering nouns, verbs, adverbs and adjectives, iii) frequency of special characters, Emoticons and disaster-related hashtags and finally iv) punctuation. Since a majority of tweets, 71%, are in English, POS annotations and sentiment analysis stick to English.

Findings. Our main findings towards the linguistic dimension are visualized in Fig. 3 and can be summarized as follows:

Informative tweets tend to:

1. be longer, on average 1.51 tokens longer than non informative ones, thus indicating that tweet length could be a suitable classification feature.
2. contain more nouns and adjectives.
3. contain URLs and the character "@" more frequently.
4. contain less positive and more negative sentiment.
5. contain less disaster related hashtags.

Non informative tweets tend to:

1. be shorter.
2. contain more verbs and adverbs.
3. contain the characters "#", "!" and "?" more frequently.
4. contain more Emoticons.
5. contain more negation terms in tweet text, such as "no", "not" or "never".
6. finish with punctuation more likely.

However, overall, *no considerable differences* between different events and different disaster types over all analyzed tweet characteristics exist.

2.4 Source Dimension

The source of a tweet was labeled by Olteanu et al. [22] and the dataset contains tweets originating from source i) business, ii) eyewitness, iii) government, iv) media, v) NGOs, iv) outsiders, and tweets not categorized, thus being grouped as "others".

Findings. Figure 4 visualizes informativeness of tweets originating from different sources. Our main findings towards the source dimension can be summarized as follows:

1. Tweets shared by "business", "media", "government" and "NGOs" tend to be *informative.*
2. Tweets shared by "eyewitness" are informative as well as non informative.
3. Tweets shared by "outsiders" and "others" tend to be *non informative.*

4. The amount of tweets shared by each source vary significantly, while 36% of all tweets originate from "media" and 33% from "outsiders", only 4% originate from "government", 1.5% from "business", 8% from "eyewitness", 3.5% from "NGO" and 14% from "others". Thus, analysis results based on these smaller classes might be not representative, especially when the amount being reduced further by considering only single events.

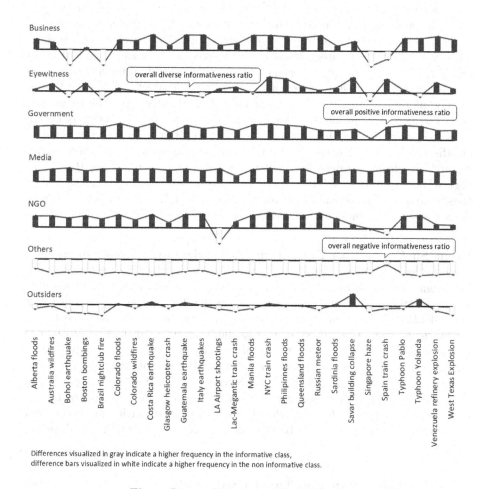

Differences visualized in gray indicate a higher frequency in the informative class,
difference bars visualized in white indicate a higher frequency in the non informative class.

Fig. 4. Source dimension - informativeness

2.5 Event Similarity

Data analysis of all four dimensions does not show any considerable differences, neither between different disaster events nor between different disaster types, which, as a consequence, supports the idea of cross-domain classification. In addition to the analysis regarding the four dimensions above, we investigate in the following how similar are tweets of events based on their characteristics in terms of a combination of temporal, spatial, linguistic and source dimension. This is realized by considering the similarity of events with respect to their tweets by applying *bottom-up hierarchical clustering*. Hierarchical clustering results are visualized in form of *cluster dendrograms*, i.e., more similar events are clustered in earlier clustering steps, thus appearing rather at the bottom of the dendrogram. In order to interpret similarity results we apply two clustering rounds: i) *event similarity based on event categorization* of Olteanu et al. [22] such as disaster type or event categorization, which serves as a baseline, and ii) *event similarity based on tweet characteristics* showing possible differences between disaster events or disaster types with respect to their tweets. While the former targets informativeness classification of individual disaster types the latter supports the idea of cross-domain classification, by considering the tweets of events intendedly disregarding the disaster type (e.g., floods or wildfires). Both rounds are discussed in more detail in the following. Agglomerative, i.e., bottom-up, hierarchical clustering used for our work is realized in R using the R package "hclust" by applying euclidean distance measure and complete linkage[4]. For those characteristics providing no information due to more than 50% of tweets are not in English, they are replaced by mean values of all other events.

Event Similarity Based on Event Categorization. In order to visualize similarity of events based on their categorization, the clustering algorithm uses the event categorization of Olteanu et al. [22], namely, duration, geographic location, category and subcategory, as well as development and spread as input and calculates based thereupon the similarity between events visualized in the cluster dendrogram of Fig. 5. As expected, the dendrogram shows certain disaster types as clusters, for instance, floods are clustered together (e.g., Philippines floods and Sardinia floods) as well as earthquakes or wildfires. With respect to interpreting cluster dendrograms, the y-axis considers the similarity between certain event clusters or events, respectively. Those events (cf. x-axis), which are clustered very early in the hierarchical progress, thus appearing at the bottom of the chart, are considered to be very similar based on hierarchical clustering results.

Event Similarity Based on Tweet Characteristics. In order to visualize similarity of events based on their tweet characteristics along the four dimensions discussed at the beginning of this section, these characteristics are used as input for the hierarchical clustering algorithms. Events are clustered considering the characteristics of tweets of events, only, intendedly disregarding the disaster type (e.g., floods or wildfires). Results of Fig. 6 show that previous clusters (e.g., the cluster comprising Colorado floods and Queensland floods), although

---

[4] https://cran.r-project.org/manuals.html.

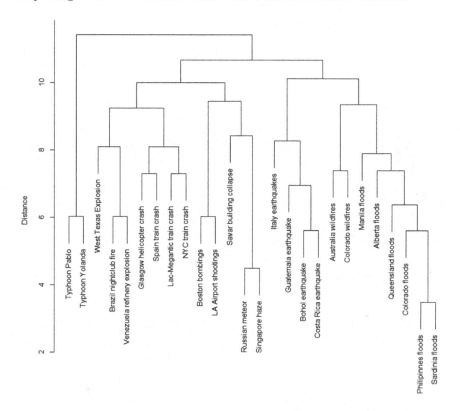

Fig. 5. Event similarity based on event categorization

stemming from the same event type, are not considered as similar based on the characteristics of their tweets. Quite the opposite, disaster events previously not considered as similar due to different disaster types, e.g., Colorado floods and Colorado wildfire, however, are considered as similar based on tweet characteristics. Hence, since events of different disaster types are considered as similar based on their tweet characteristics, thus further supporting the idea of cross-domain informativeness classification.

Interpreting Hierarchical Clustering Results. First of all, comparing results of both clustering rounds above shows differences, meaning that the disaster type is not as relevant for informativeness classification of tweets, since clustering based on tweet characteristics does not show clusters with respect to the disaster types. Hence, the similarity of events based on tweet characteristics can be interpreted as people's online communication behavior in disaster events, which is, based on results so far, not distinguishing between different disaster types. For instance, the events Colorado floods and Colorado wildfires are considered to be similar based on their tweet characteristics, although their disaster type is not the same. In other words, whether a tweet is informative or

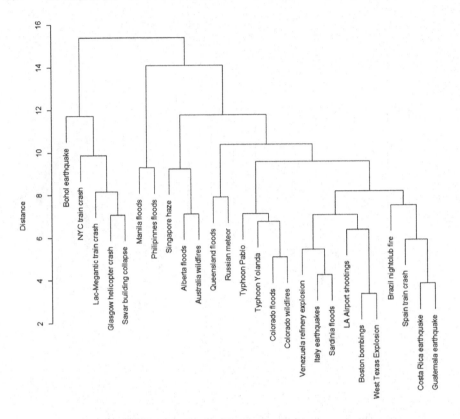

Fig. 6. Event similarity based on tweet characteristics

not depends on certain characteristics, which are apparently not considerably different for different disaster types.

2.6   Tweet Similarity

Hierarchical clustering visualizes event similarity rather than the similarity of individual tweets. In order to visualize similarity of individual tweets, thereby showing up possible clusters with respect to disaster types, we additionally visualize all tweets in a scatter plot. Thereby, Principal Component Analysis (PCA) [1] implemented in Python's Scikit-Learn library[5] was used to visualize d-dimensional information of tweets, i.e., all tweet characteristics described along the four dimensions discussed at the beginning of this section, in a two dimensional space. Results of Fig. 7 support once again the idea of cross-domain informativeness classification of tweets, since no clusters related to disaster types are recognizable, i.e., tweets from all events are distributed over the entire plot.

---

[5] https://scikit-learn.org/stable/modules/generated/sklearn.decomposition.PCA.
html.

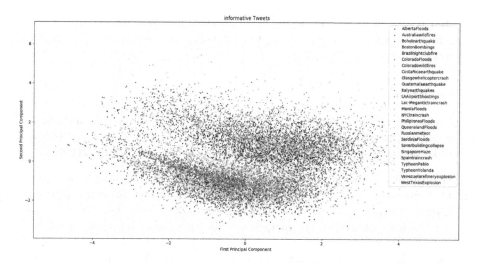

Fig. 7. Tweet similarity visualized by applying PCA

2.7 Summary of Outcomes of Our Data Analysis

Our main findings along all four dimensions as well as findings along event and tweet similarity can be summarized in *two outcomes*: Firstly, analysis of temporal, linguistic and source dimensions uncovers tweet characteristics having impact on informativeness, thus being candidates to be used as features, whereas analysis of spatial dimension (per continent) shows no significant correlations with respect to informativeness. Secondly, the fact that analysis does not show significant differences with respect to disaster types and furthermore does not show significant similarities within same disaster types, supports the hypotheses to train an accurate cross-domain classifier cross the 13 different disaster types available in our dataset.

3 Classification Approach

Based on the analysis presented in the previous section suggesting that different disaster types do not show significant differences in terms of informativeness and taking into account indication of informativeness of certain characteristics, in this section we elaborate on a classification approach visualized in Fig. 8, which can be used on new events of various disaster types while being at least as accurate in informativeness classification as disaster type specific ones. Thereby, the proposed classification approach resembles state-of-the-art approaches of related work (cf. Table 4 in Sect. 5) comprising three phases, namely i) training phase, ii) testing phase, and iii) online classification phase. Firstly, during the training phase, an appropriate training set is used to train a classifier. Secondly, during testing phase, this classifier is evaluated by using appropriate unseen test sets. Finally, the trained classifier can be applied to an online stream of disaster related tweets in order to classify in informative and non informative tweets. Our implementation was realized in Python using the frequently used Machine Learning library *Scikit-Learn* [23].

3.1 Dataset

Split Dataset into Training Set and Test Set. In order to evaluate which classifier, a cross-domain one or an event specific one, is more suited for new events, we consider three different types of experiments regarding the disaster type:

1. **In-domain**: training and test data belong to the *same* disaster type.
2. **Out-domain**: training and test data belong to *different* disaster types.
3. **Cross-domain**: training set consists of tweets of *various* disaster types, the test set is of a disaster type included or not included in the training set.

Sampling. To cope with the fact that our dataset is unbalanced in that the total amount of informative tweets and the total amount of non informative tweets is not equal in the dataset, and some classification algorithms are vulnerable with respect to unbalanced data, we use an oversampling strategy [23] by replicating samples from the smaller class. In addition, sampling allows to reduce the size of the training set to evaluate the impact of training size on classification accuracy.

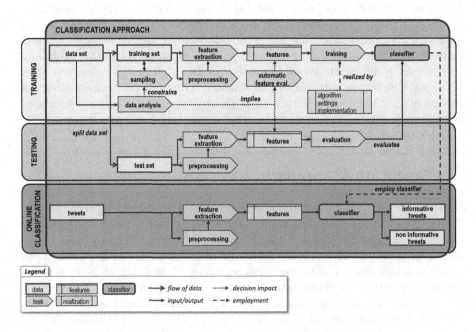

Fig. 8. Schematic representation of classification approach

3.2 Preprocessing and Feature Extraction

In order to extract features, depending on their respective kinds, preprocessing steps comprise removing stopwords, tokenization, stemming and POS annotation, which are realized on basis of the NLTK library[6], as well as language detection on basis of the Langdetect[7] library and sentiment determination on basis of the TextBlob[8] library. Which features to use is based on findings during data analysis, discussed in the previous Sect. 2. In the following, these features and their extraction process from tweets are described in more detail. Finally, our classification approach ended up with a set of features listed in Table 2.

Table 2. Set of features - grouped by analysis dimensions

Nr.	Dimension	Feature	Data type
1	Temporal	Response time	Integer
2–5		Probability of language EN, ES, TL, PT	Float [0,1]
6		Number of tokens	Integer
7–10		Characters "#", "?", "!", "@"	Integer
11		Links (URLs)	Integer
12–15		POS: nouns, verbs, adjectives, adverbs	Integer
16–17	Linguistic	Positive/negative Emoticons	Integer
18		Sentiment polarity	Float [−1,1]
19		Sentiment subjectivity	Float [0,1]
20		Disaster related Hashtags	Integer
21		Negation terms	Integer
22		Tweet finishes with punctuation	Binary
23		Media	Binary
24		Business	Binary
25		NGO	Binary
26	Source	Government	Binary
27		Eyewitness	Binary
28		Others	Binary
29		Outsiders	Binary

Response Time. The response time of a tweet, i.e., the time span from the beginning of a disaster to the point in time when a certain tweet is shared online,

---

[6] https://www.nltk.org/.
[7] https://pypi.python.org/pypi/langdetect.
[8] http://textblob.readthedocs.io/en/dev/index.html.

contains important information with respect to informativeness. The response time is calculated by the formula below and is measured in days.

$$featureResponseTime = tweetPostDate - eventStarted \qquad (1)$$

Probability of Language EN, ES, TL, PT. Naturally, language information is crucial in combination with other features due to the reason that some features are based on the language of the tweet. Top four languages prevalent in tweets of the dataset are English (EN), Spanish (ES), Tagalog (TL), and Portuguese (PT), together comprising more than 90% of all tweets. The language of each tweet is determined using the Python library Langdetect, a language detection tool ported from Google's language-detection determining the probability of languages used for a certain tweet. For our approach we use the language detection for the languages EN, ES, TL, and PT, resulting in four features for classification.

Number of Tokens. After preprocessing the tweet text, more precisely, removing stopwords, tokenization, and stemming on basis of the NLTK library, the number of remaining tokens of the tweet's text is determined, representing another feature for classification.

Special Characters. Without preprocessing the tweet text, i.e., from tweet raw data, the number of characters occurring in the tweet text is determined. Each of the four characters ("#", "?", "!", "@") result in one feature for classification, thus four features in total, expressing how often a special character occur in tweet text.

Links (URLs). The number of URLs referring to web-links is determined simply by searching "http" in tweet raw data. This results in one feature for classification expressing the number of links in a tweet text.

POS: Nouns, Verbs, Adjectives, Adverbs. Part-of-Speech (POS) annotation is based on the Python NLTK library. More precisely the POS tagger function "pos_tag" is used based on appropriately reprocessed tweets (i.e., stopword removal, tokenization, and stemming), resulting in four different features for classification, namely i) number of nouns, ii) number of verbs, iii) number of adjectives, and iv) number of adverbs.

Positive/Negative Emoticons. The number of Emoticons occurring in tweet text are determined by searching for particular text sequences, such as ":-)" or ":-(". Thereby, variations of positive and negative Emoticons are considered resulting in two features for classification.

Sentiment Polarity/Subjectivity. In addition to determining the number of Emoticons, only, sentiment analysis tools use a variety of techniques to assess the sentiment of a text. Sentiment analysis of our approach is based on the Python library TextBlob providing a sentiment analysis module, which uses a pattern analyzer to estimate the sentiment in terms of polarity and subjectivity of a particular text. The resulting two features for classification represent the sentiment of a tweet text. The first feature expresses the sentiment polarity, i.e, whether a tweet contains more likely positive or negative sentiment. The second feature expresses the subjectivity of the sentiment, i.e., whether the sentiment polarity value is rather subjective or objective.

Disaster Related Hashtags. Disaster related Hashtags, such as #HurricanSandy, are used by twitter users to refer to certain disaster events, many of them occurring in tweet text. For our approach we determine the number of disaster related Hashtags by matching already known disaster related Hashtags given in the CrisisLexT26 dataset with Hashtags in the tweet text. The number of matches is represented as a feature value for classification.

Negations Terms. The number of negation terms listed below is used as another feature for classification. Its value represents the number of negation terms occurring in the tweet text, covering the terms "not", "none", "neither", "never", "no one", "nobody", "nor", "nothing", "nowhere", "does not", "did not", and "f*ck" (profane words are spelled with "*" to replace one letter).

Tweet Finishes with Punctuation. This binary feature for classification distinguishes tweets finishing with punctuation, more precisely with the characters ".", "!" or "?", from tweets finishing without punctuation.

Source. Not least since the source dimension showed considerable differences regarding informativeness in data analysis, six binary features indicating a tweet's source are used for classification: i) Media, ii) Business, iii) NGO, iv) Government, v) Eyewitness, and vi) Outsiders and (vii) Others, covering source-agnostic tweets.

3.3   Automatic Feature Evaluation

Whether the features and their settings described above are suitable for the final informativeness classification approach of our work is additionally evaluated by automatic feature evaluation metrics. Several metrics are provided by the Scikit-Learn framework. One of them is the "mutual information classification"[9], which measures the dependency between the feature and a certain class label. Higher

---

[9] https://scikit-learn.org/stable/modules/generated/sklearn.feature_selection. mutual_info_classif.html.

values mean higher dependency, thus mean higher importance of this particular feature for classification. Results of automatic feature evaluation are listed by decreasing dependency order in Table 3 and visualized in Fig. 9 indicating that the first feature "Media" of dimension source is the most important one for informativeness classification due to showing highest dependency to class. The following feature "Links (URLs)" has the second highest dependency to class and is therefore the second most important feature for informativeness classification. However, considering these two most important features (cf. Fig. 9), only, it is noticeable that the feature "Media" shows a much higher dependency than all others meaning that the source "Media" is by far the most relevant indicator for informativeness of tweets. Besides this fact of the highly important feature "Media" for classification, we can identify two further groups of features with respect to their dependency to class results. We identify features starting from "Links (URLs)", i.e., rank 2, to feature "Language EN", i.e., rank 11, as very important features for classification due to a dependency to class value higher then 0.02, however, showing continuously decreasing dependency to class. Finally, all other features, i.e., from rank 12 to the end, we still identify as important for classification due to showing dependency to class, however, their differences regarding dependency to class between individual features are small. With respect to our analysis dimensions we can conclude that all of the three dimensions namely source, linguistic, and temporal are highly relevant for informativeness classification since features of all of these dimensions are within the top eight most relevant features for informativeness classification.

3.4   Training and Testing the Classifier

Training and Testing. In order to simulate a real scenario where tweets are classified with respect to informativeness on new events, our trained classifiers are tested on unseen disaster events. To compare their classification results, we test classifiers on the *same* disaster event, and employ classification accuracy as basis of our evaluation, i.e., how many tweets related to the total number of classified tweets are classified as informative or non informative correctly.

Algorithm, Settings and Implementation. State-of-the-art informativeness classification employ standard algorithms, like Support Vector Machines (SVM) [5,14,20,24,31], Naive Bayes classification, Maximum Entropy Models [24,31,33] or Random Forest classification [2,12,19] as well as deep learning [21], using a Convolutional Neural Network (CNN). Classification algorithms provided by Scikit-Learn, such as SVM, Naive Bayes, AdaBoost (an ensemble method), Random Forest, and a Multilayer Perceptron (a neural network) have been used for the presented experiments. However, experiments showed that SVM using an RBF kernel by applying Scikit-Learn default settings (C = 1.0, gamma = 'auto') work best out of all other algorithms being therefore the first choice for all of our experiments.

Table 3. Mutual information classification - on average over all experiments

	Dimension	Feature	Dependency to class
1	Source	Media	0.134248132
2	Linguistic	Links (URLs)	0.058554753
3	Source	Outsiders	0.053998574
4	Linguistic	Sentiment objectivity	0.048314646
5	Linguistic	POS nouns	0.046057527
6	Linguistic	Sentiment subjectivity	0.038255962
7	Linguistic	Characters "#"	0.037680616
8	Temporal	Response time	0.031614661
9	Linguistic	Number of tokens	0.025143307
10	Linguistic	Disaster related Hashtags	0.022630527
11	Linguistic	Language EN	0.021173189
12	Linguistic	POS adjective	0.016483257
13	Linguistic	Characters "!"	0.015189369
14	Linguistic	Characters "?"	0.015153933
15	Linguistic	Characters "@"	0.015095157
16	Linguistic	POS adverb	0.012776012
17	Linguistic	Language TL	0.012484242
18	Source	Eyewitness	0.012277692
19	Source	Government	0.011127693
20	Linguistic	POS verbs	0.010531956
21	Linguistic	Sentence finishes with punctuation	0.009901169
22	Source	NGO	0.009665882
23	Linguistic	Language ES	0.008910637
24	Linguistic	Language PT	0.008864438
25	Source	Business	0.006817192
26	Linguistic	Negation terms	0.006486926
27	Linguistic	Positive Emoticons	0.006133980
28	Linguistic	Negative Emoticons	0.005494237

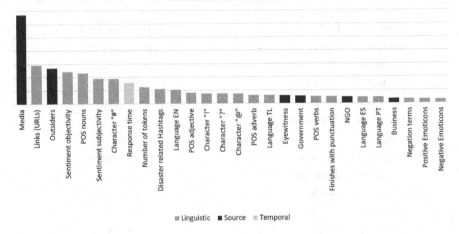

Fig. 9. Visualization of mutual informativeness classification (cf. Table 2)

4   Evaluation

Evaluation Dimensions. For a systematic evaluation of our classification approach, described in the previous section, we define two orthogonal dimensions for our experiments:

1. **Event specifity**: To clarify the question whether disaster type specific classifiers or more generic cross-domain classifiers are more beneficial, we distinguish training data containing tweets of only one type of disaster (i.e., deep event specifity) from training data including tweets of multiple different types of disasters (i.e., broad event specifity).
2. **Training size**: To clarify the impact of sample size on classifier performance in relation to event specifity, we distinguish the amount of tweets used for training from 1K (small training size) to 28K (large training size).

Experiment Categorization. Based on our evaluation dimensions, we categorize our experiments into four groups as shown in Fig. 10. Experiments using a large amount of tweets for training and a deeper event specificity, however, had to be ruled out, since a comparable amount of tweets of the same disaster type is not available in the CrisisLexT26 data set. By comparing classification results based on the three remaining groups, we are able to show the impact of the disaster type, i.e., event specificity, as well as the impact of the amount of tweets used for training the classifier on classification accuracy. Thus, we can show, how accurate our cross-domain classifier performs on various events of different disaster types against classifiers trained on events of the same disaster type (e.g., trained on an earthquake event "Guatemala earthquake" and tested on an earthquake event "Costa Rica earthquake").

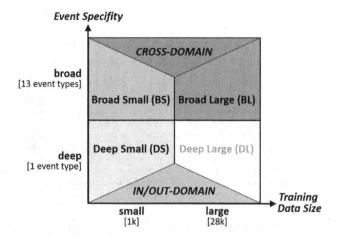

Fig. 10. Evaluation dimensions and experiment categorization

4.1 Deep—Small Experiments (DS)

In order to get an evaluation baseline for classification, first we consider systematic in-domain and out-domain classification experiments by using all possible train/test-set combinations visualized in Fig. 11. Regarding training data size, this kind of experiments use a rather small amount of approximately 1K tweets to train the classifier, since limited by the available CrisisLexT26 dataset. For those disaster types in our dataset which contain data about more than one event, disaster events of the same disaster type are used for training and testing the classifier, i.e., in-domain classification. For instance, for disaster type "floods", the event Alberta floods is used to train the classifier and the event Sardinia floods is used to test the classifier. In total, the CrisisLexT26 dataset allows 55 in-domain experiments, by using all possible train/test-set combinations of disaster events of the same disaster type, considering those types identified by Olteanu et al. [22]. Using all possible train/test-set combinations of disaster events of different disaster types result in additional 495 out-domain experiments. Figure 12 (cf. chart lines "DS") shows the average classification accuracy of all 650 in- and out-domain experiments with respect to one particular disaster event used for testing the classifier. In-domain experiments result on average in an informativeness classification accuracy of 75% (including a standard deviation of 5%). Out of 55 in-domain experiments, the best classification accuracy of 88% achieved the Costa Rica earthquake using the Guatemala Earthquake as training event, which might be due to obvious similarities between these two events. The worst result of 58% results from the event Philippines flood using Sardinia floods as training set. In order to verify our hypothesis that a cross-domain classifier leads to at least as accurate informativeness classification as an in-domain classifier, i.e., a more specific one, the mentioned classification results serve as a baseline.

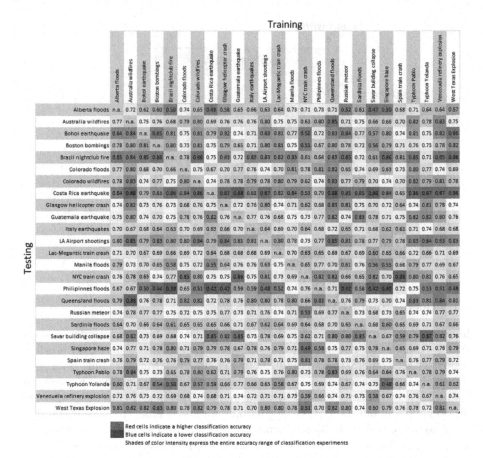

Fig. 11. In-domain and out-domain classification accuracy results

4.2 Broad—Small Experiments (BS)

In order to eliminate the impact of training data size on classification accuracy when comparing results against DS experiments, in our second group of experiments we use the same amount of tweets as before, for training. We sample these 1K tweets out of all disaster events, excluding the one disaster event used for testing the classifier, to address a broad event specificity. In general, our cross-domain classification experiments follows a "leave one out" strategy, 25 disaster events are used for training and the remaining 26th event was used for testing. The average classification accuracy over all 26 experiments is 79% (including a standard deviation of 7%), which is 4% higher than the average of in-domain experiments (cf. Fig. 12, chart lines "BS"). In other words, a *classifier trained on random sampled tweets* from various disaster events of different disaster types *achieves a 4% higher informativeness classification accuracy* than using a classifier *trained on the same disaster type* as the actual disaster is, even *using the same size* of tweets for training.

4.3   Broad—Large Experiments (BL)

Since using tweets of different disaster events of different disasters types for training lead to more accurate classification on average than in-domain training, in the third group of our experiments we want to figure out the impact of the training size on classification accuracy. By applying a "leave one out" strategy, our 26 experiments use all available 28K tweets, again excluding those used for testing, to train the classifier. Classification results are visualized in Fig. 12, chart lines "BL". On average, classification accuracy over all cross-domain experiments is 80% (including a standard deviation of 7%), which is significantly higher compared to in- and out-domain experiments. Comparing the average classification results using 28K tweets for training against using 1K tweets, show a slightly, 1%, higher classification accuracy. Thus, the interesting finding here is that the size of the dataset used for training seems not to be primary relevant for accurate classification of informativeness.

4.4   Interpretation of Evaluation

To sum up, experimentation results allow us the following *conclusions*:

1. Using a classifier trained on *various* events cross different types of disasters outperforms in 23 cases out of 26 (cf. red diamonds in Fig. 12) more specific classifiers, trained on the same disaster type, in classification accuracy of 4% (avg.). This even with the same amount of tweets in the training data.
2. Using a more specifically trained classifier may tend to overfit and therefore leads to less accurate informativeness classification of unseen disaster events.
3. Increasing the amount of training samples on average lead to slightly more accurate classification only (cf. Fig. 12).

4.5   Classification Performance

Finally, when considering informativeness classification results of other closely related approaches, which are discussed in more detail in the following related work section, as a baseline, results show that the cross-domain classifier presented in the current work achieved a higher accuracy on average as well as regarding best accuracy values (cf. Table 4 in Sect. 5). In particular, compared to cross-domain trained classifiers of [2] as well as to the in-domain classifier of [19], our cross-domain classifier is 4% (avg.) more accurate in informativeness classification, and compared to [12] 1.3% more accurate. In the current work, the best average accuracy is 80% (including a standard deviation of 7%) using cross-domain training over all events, best accuracy classification result is 89%, which is compared against [12] an accuracy improvement of 2% and 12% compared against [19]. The worst result was classifying Philippines flood with an accuracy of 62%.

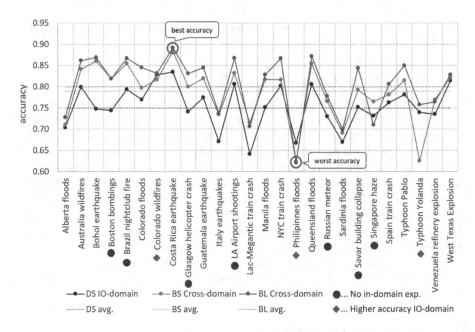

Fig. 12. Classification accuracy results (Color figure online)

5   Related Work

Informativeness Classification. Informativeness is a broadly discussed concept in literature and applied in various areas like news articles [18], web documents [9] and linguistic sciences [16,37], to mention just a few. In the disaster domain, informativeness classification is one early step in processing information from social media. While few research thereby focuses on the informativeness classification task [2,12,19], other work take that for granted and focus on more specific classification tasks like damage assessment [5] or develop platforms [4] and frameworks [3,26], which support disaster management as a whole. While many approaches for informativeness classification focus on particular disaster types [2,5,19], only few works, however, address classification cross a variety of disaster types [13,14,25] and also in other application areas like e.g. news [6] cross-domain approaches are sparse. Table 4 shows a summary of closely related work, sorted firstly with respect to the classification task, secondly concerning the dataset used and thirdly regarding other related aspects, such as used features or cross-domain training. After discussing related work with respect to systematic manual analysis of disaster data, in the following, the main differences to other closely related classification approaches are factored out.

Manual Disaster Data Analysis. While many research focus on feature engineering, i.e., inventing and evaluating new features, [13,14,19] or classification itself [24,31,33], only few research deal with disaster data analysis *with respect to informativeness.* Acerbo and Rossi investigate "common patterns" inside informative and non informative tweets, which is similar to the data analysis of our work, yet, they focus on similarities and differences in words. A statistical analysis of data with respect to informativeness is done by Lloret and Palomar [18] where they present linguistic features which "an informative tweet should have in order to be informative", focusing, however, on the news domain. Ning et al. [21] present an analysis of six disasters, to identify linguistic, sentimental and emotional features. In contrast to our work, they address "relatedness" rather than "informativeness" of tweets. The manual disaster data analysis of our work is based on the results of Olteanu et al. [22], additionally, on top of that we focus on a detailed analysis with respect to informativeness.

Classification Approaches. Considering closely related informativeness classification approaches, Acerbo and Rossi [2] base their work on a subset of the CrisisLexT26. In contrast to our work, their goal was not to learn a cross-domain classifier, rather they focus on a novel text metric to use as additional feature for classification. Moreover, their dataset contains only three disaster types: floods, earthquakes and fires. Closely related with respect to the number of cross-domain experiments is the work of Imran et al. [12], which is also based on a subset of the CrisisLexT26, yet, they only use two disaster types, namely floods and earthquakes. Longhini et al. [19] present a "language-agnostic model" for informativeness classification and show the impact of a new feature "source", indicating hardware for communication, e.g. mobile or not. Cross-domain classification, however, is not part of their work. Considerable more disaster types as all previously mentioned approaches are used by Khare et al. [14]. As in our work, they run cross-domain experiments, yet their classification task is "relatedness" (whether a tweet is related to a disaster event or not) instead of "informativeness". Most closely related to ours is the work of [13] where they use the entire CrisisLexT26 dataset to address cross-domain classification, focusing, however, on "relatedness", not on "informativeness". The work of [5] is closely related with respect to features used for classification like number of tokens, "@"-symbols, hashtags, punctuation, Emoticons and sentiment. Additionally, they address cross-domain classification by using four disaster events of two disaster types in Italy. They do not focus, however, on informativeness classification but rather on "damage assessments" of Italian tweets. Closely related with respect to cross-domain classification are the works of Li et al. [17] and Imran et al. [11], yet they consider two different disaster types, only.

Table 4. Overview on informativeness classification approaches.

Approach	Goal	Dataset	Language	No. tweets	No. disasters	No. types	Linguistic	Temporal	Spatial	Algorithm	In-domain	Out-domain	Cross-domain	Best accuracy	Avg. accuracy
Acerbo & Rossi [2]	Informativeness	CrisisLexT26	EN,ES,TL	36.0K	12	3	✓	✓		Random Forest			✓	n.a.ᵃ	76.0%ᵇ
Imran et al. [12]	Informativeness	CrisisLexT26	EN,ES,TL	11.7K	11	2	✓	✓		Random Forest	✓	✓	✓	86.0%ᵇ	78.7%ᵇ
Longhini et al. [19]	Informativeness	CrisisLexT26	EN,ES,TL	12.9K	12	3	✓	✓		Random Forest	✓			75.9%	71.8%
Khare et al. [14]	Relatedness	CrisisLexT26	EN	3.2K	9	7	✓			SVM			✓	82.3%ᵇ	73.1%ᵇ
Khare et al. [13]	Relatedness	CrisisLexT26	EN,ES,TL,PT	32.0K	26	11	✓			SVM		✓	✓	n.a.ᵃ	n.a.ᵃ
Cresci et al. [5]	Damage assessment	own	IT	5.6K	4	2	✓			SVM	✓	✓	✓	70.0%ᵇᶜ	44.0%ᵇᶜ
Ning et al. [21]	Relatedness	CrisisLexT6	EN	32.5K	6	5	✓		✓	CNN			✓	87.1%ᵇ	n.a.ᵃ
Li et al. [17]	Relatedness and others	own	EN	2.7K	2	2	✓			Naive Bayes			✓	n.a.ᵃ	n.a.ᵃ
Imran et al. [11]	Informativeness and others	own	EN	6.4K	2	2	✓			Naive Bayes	✓			87.1%	n.a.ᵃ

ᵃinformation not available, ᵇcross-domain results, ᶜclassification in three classes

6   Summary and Outlook

To sum up, our work presented in this article comprises the following three con-tributions: First, in order to better understand the underlying social media data source available in disaster situations, an in-depth analysis of existing Twitter data on 26 different disaster events is provided along four different dimensions covering temporal, spatial, linguistic and source. Second, based thereupon, a cross-domain informativeness classifier is proposed being not focused on specific disaster types but rather being able to provide for classifications across different types. Third, the applicability of the cross-domain classifier is demonstrated, showing the high accuracy of our approach compared to other disaster type specific approaches. More precisely, our proposed cross-domain trained classifier shows following benefits:

1. It is usable on various events of various types of disasters so that a single classifier is applicable for any event.
2. It achieves 4% (avg.) higher classification accuracy than disaster-type specific classifiers using the same size of training data.
3. It increases the amount of available training data since being not limited to one type of disaster.

Based on our contributions, several lines of future research can be identified. First, however, evaluation of classification is based on the 26 disaster events and 13 disaster types included in the CrisisLexT26 data set, only. Hence, results might be different when classifying new disaster types. In addition, since a major-ity of tweets in the CrisisLexT26 data set are in English, it can be expected that classification is not as accurate in classifying tweets of other languages thus limiting the classifiers applicability, also confirmed by the related work of [15]. Classifying tweets of other languages requires to adjust linguistic features (e.g. POS, sentiment) to particular languages. Moreover, the CrisisLexT26 data set contained tweets stemming from the years 2012 to 2013. Yet, communication of people might change over years and thus informativeness classification accuracy using classifiers trained on past events happening years ago might decrease over time. User-related information like user meta data or geo-location are expected to further improve the overall quality of classification, but has not been considered in the current work since not being included in the CrisisLexT26 dataset. In our work we use mainly linguistic, emotional and sentimental features. Experiments including additional (semantic) features as well as geo-location information like spatial relationship of tweets [32], which could help increase the accuracy of the classifier, is subject to work. Another line of intended research is the consideration of other languages, disaster data sets and finally also the expansion of our evaluation towards other domains like the classification of news articles or web documents realizing a domain-generic informativeness classification approach.

References

1. Abdi, H., Williams, L.J.: Principal component analysis. Wiley Interdisc. Rev. Comput. Stat. **2**(4), 433–459 (2010)
2. Acerbo, F., Rossi, C.: Filtering informative tweets during emergencies: a machine learning approach. In: Proceedings of the 1st CoNEXT Workshop on ICT Tools for Emergency Networks and Disaster Relief, I-TENDER 2017, pp. 1–6. ACM, New York (2017)
3. Avvenuti, M., Cimino, M.G.C.A., Cresci, S., Marchetti, A., Tesconi, M.: A framework for detecting unfolding emergencies using humans as sensors. SpringerPlus **5**(1), 1–23 (2016). https://doi.org/10.1186/s40064-016-1674-y
4. Cameron, M., Power, R., Robinson, B., Yin, J.: Emergency situation awareness from twitter for crisis management. In: Proceedings of the 21st International Conference on World Wide Web, WWW 2012, pp. 695–698. ACM, New York (2012)
5. Cresci, S., Tesconi, M., Cimino, A., Dell'Orletta, F.: A linguistically-driven approach to cross-event damage assessment of natural disasters from social media messages. In: Proceedings of the 24th International Conference on World Wide Web, WWW 2015, pp. 1195–1200. ACM (2015)
6. Dai, W., Xue, G., Yang, Q., Yu, Y.: Transferring Naive Bayes classifiers for text classification. In: Proceedings of the 22nd International Conference on Association for the Advancement of Artificial Intelligence, AAAI 2007, vol. 7, pp. 540–545 (2007)
7. Derczynski, L., Meesters, K., Bontcheva, K., Maynard, D.: Helping crisis responders find the informative needle in the tweet haystack. arXiv preprint arXiv:1801.09633 (2018)
8. Girtelschmid, S., Salfinger, A., Pröll, B., Retschitzegger, W., Schwinger, W.: Near real-time detection of crisis situations. In: Proceedings of 39th International Convention on Information and Communication Technology, Electronics and Microelectronics, MIPRO 2016, pp. 247–252. IEEE (2016)
9. Horn, C., Zhila, A., Gelbukh, A., Kern, R., Lex, E.: Using factual density to measure informativeness of web documents. In: Proceedings of the 19th Nordic Conference of Computational Linguistics, NODALIDA 2013, pp. 227–238 (2013)
10. Imran, M., Castillo, C., Diaz, F., Vieweg, S.: Processing social media messages in mass emergency: a survey. ACM Comput. Surv. (CSUR) **47**, 1–38 (2015)

11. Imran, M., Elbassuoni, S., Castillo, C., Diaz, F., Meier, P.: Extracting information nuggets from disaster-related messages in social media. In: Proceedings of the 10th Conference for Information Systems for Crisis Response and Management, ISCRAM 2013 (2013)
12. Imran, M., Mitra, P., Srivastava, J.: Cross-language domain adaptation for classifying crisis-related short messages. arXiv preprint arXiv:1602.05388 (2016)
13. Khare, P., Burel, G., Alani, H.: Classifying crises-information relevancy with semantics. In: Gangemi, A., et al. (eds.) ESWC 2018. LNCS, vol. 10843, pp. 367–383. Springer, Cham (2018). https://doi.org/10.1007/978-3-319-93417-4_24
14. Khare, P., Fernandez, M., Alani, H.: Statistical semantic classification of crisis information. In: 1st workshop of Hybrid Statistical Semantic Understanding and Emerging Semantics (HSSUES), 16th International Semantic Web Conference (2017)
15. Khare, P., Burel, G., Maynard, D., Alani, H.: Cross-lingual classification of crisis data. In: Vrandečić, D.D., et al. (eds.) ISWC 2018. LNCS, vol. 11136, pp. 617–633. Springer, Cham (2018). https://doi.org/10.1007/978-3-030-00671-6_36
16. Kireyev, K.: Semantic-based estimation of term informativeness. In: Proceedings of the 2009 Conference of the North American Chapter of the Association for Computational Linguistics, NAACL-HLT 2009, pp. 530–538. Association for Computational Linguistics (2009)
17. Li, H., et al.: Twitter mining for disaster response: a domain adaptation approach. In: Proceedings of the 12th Conference for Information Systems for Crisis Response and Management, ISCRAM 2015 (2015)
18. Lloret, E., Palomar, M.: Analysing and evaluating the task of automatic tweet generation: knowledge to business. Comput. Ind. **78**, 3–15 (2016)
19. Longhini, J., Rossi, C., Casetti, C., Angaramo, F.: A language-agnostic approach to exact informative tweets during emergency situations. In: International Conference on Big Data, Big Data 2017, pp. 3475–3739. IEEE (2017)
20. Mohammad, S., Kiritchenko, S., Zhu, X.: NRC-Canada: building the state-of-the-art in sentiment analysis of tweets. arXiv preprint arXiv:1308.6242 (2013)
21. Ning, X., Yao, L., Wang, X., Benatallah, B.: Calling for response: automatically distinguishing situation-aware tweets during crises. In: Cong, G., Peng, W.-C., Zhang, W.E., Li, C., Sun, A. (eds.) ADMA 2017. LNCS (LNAI), vol. 10604, pp. 195–208. Springer, Cham (2017). https://doi.org/10.1007/978-3-319-69179-4_14
22. Olteanu, A., Vieweg, S., Castillo, C.: What to expect when the unexpected happens: social media communications across crises. In: Proceedings of the 18th ACM Conference on Computer Supported Cooperative Work & Social Computing, CSCW 2015, pp. 994–1009. ACM (2015)
23. Pedregosa, F., et al.: Scikit-learn: machine learning in Python. J. Mach. Learn. Res. **12**, 2825–2830 (2011)
24. Pekar, V., Binner, J., Najafi, H., Hale, C.: Selecting classification features for detection of mass emergency events on social media. In: Proceedings of the 15th International Conference on Security and Management, SAM 2016, The Steering Committee of The World Congress in Computer Science, Computer Engineering and Applied Computing (WorldComp), p. 192 (2016)
25. Pekar, V., Binner, J., Najafi, H., Hale, C., Schmidt, V.: Early detection of heterogeneous disaster events using social media. J. Assoc. Inf. Sci. Technol. **71**, 43–54 (2020)

26. Ren, X., et al.: CoType: joint extraction of typed entities and relations with knowledge bases. In: Proceedings of the 26th International Conference on World Wide Web, WWW 2017, International World Wide Web Conference on Steering Committee, pp. 1015–1024 (2017)
27. Rossi, C., et al.: Early detection and information extraction for weather-induced floods using social media streams. Int. J. Disaster Risk Reduct. **30**, 145–157 (2018)
28. Salfinger, A.: Staying aware in an evolving world. Ph.D. thesis, Johannes Kepler University of Linz (2016)
29. Salfinger, A., Salfinger, C., Pröll, B., Retschitzegger, W., Schwinger, W.: Pinpointing the eye of the hurricane-creating a gold-standard corpus for situative geo-coding of crisis tweets based on linked open data. In: LDL 2016 5th Workshop on Linked Data in Linguistics: Managing, Building and Using Linked Language Resources, p. 27 (2016)
30. Salfinger, A., Schwinger, W., Retschitzegger, W., Pröll, B.: Mining the disaster hotspots-situation-adaptive crowd knowledge extraction for crisis management. In: Proceedings of the 2016 Multi-Disciplinary International Conference on Cognitive Methods in Situation Awareness and Decision Support, CogSIMA 2016, pp. 212–218. IEEE (2016)
31. Stowe, K., Paul, M., Palmer, M., Palen, L., Anderson, K.: Identifying and categorizing disaster-related tweets. In: Proceedings of The 4th International Workshop on Natural Language Processing for Social Media, pp. 1–6 (2016)
32. Tsuchida, T., Kato, D., Endo, M., Hirota, M., Araki, T., Ishikawa, H.: Analyzing Relationship of words using biased LexRank from geotagged tweets. In: Proceedings of the 9th International Conference on Management of Digital Ecosystems, MEDES 2017, pp. 42–49. ACM, New York (2017)
33. Verma, S., et al.: Natural language processing to the rescue? Extracting "situational awareness" tweets during mass emergency. In: Proceedings of the 5th Conference on Weblogs and Social Media, ICWSM 2011 (2011)
34. Vieweg, S.: Situational awareness in mass emergency: a behavioral and linguistic analysis of microblogged communications. Ph.D. thesis, University of Colorado at Boulder (2012)
35. Vieweg, S., Hughes, A.L., Starbird, K., Palen, L.: Microblogging during two natural hazards events: what twitter may contribute to situational awareness. In: Proceedings of the Conference on Human Factors in Computing Systems, CHI 2010, pp. 1079–1088. ACM (2010)
36. Wong, B., Kit, C.: Comparative evaluation of term informativeness measures for machine translation evaluation metrics. In: Proceedings of the 13th Conference of Machine Translation Summit, vol. 2011, pp. 537–544 (2011)
37. Wu, Z., Giles, C.: Measuring term informativeness in context. In: Proceedings of the 2013 Conference of the North American Chapter of the Association for Computational Linguistics: Human Language Technologies, NAACL-HLT 2013, pp. 259–269 (2013)

COTILES: Leveraging Content and Structure for Evolutionary Community Detection

Nikolaos Sachpenderis, Georgia Koloniari$^{(\boxtimes)}$, and Alexandros Karakasidis

Applied Informatics Department, University of Macedonia,
Thessaloniki 54636, Greece
{sachpenderis,gkoloniari,a.karakasidis}@uom.edu.gr

Abstract. Most community detection algorithms for online social networks rely solely either on the structure of the network, or on its contents. Both extremes ignore valuable information that influences cluster formation. We propose COTILES, an evolutionary community detection algorithm, that leverages both structural and content-based criteria so as to derive densely connected communities with similar contents. Specifically, we extend a fast online structural community detection algorithm by applying additional content-based constraints. We also further explore the effect of structure and content-based criteria on the clustering result by introducing three tunable variations of COTILES that either tighten or relax these criteria. Through our experimental evaluation, we show that the proposed method derives more cohesive communities compared to the original structural one, and highlight when the proposed variations should be deployed.

Keywords: Community detection · Social networks · Labeled communities · Evolutionary clustering

1 Introduction

Community detection and analysis in social networks is of particular interest as it finds many applications such as marketing and advertising, while it is also very demanding as social networks are massive and highly dynamic. They constantly evolve both with regards to their structure as nodes join and leave the network, form new links and remove old ones, and also with regards to their contents as user interests and content of interactions change over time as well.

As communities in social networks are usually defined as groups of nodes that are densely connected among themselves while sharing less connections with other members of the network, most techniques relegate the problem of community detection to a graph clustering task where only the structure of the graph is taken into account to detect dense subgraphs. However, the content of the social graph that can be represented by edges' or nodes' labels describing

© Springer-Verlag GmbH Germany, part of Springer Nature 2020
A. Hameurlain et al. (Eds.) TLDKS XLV, LNCS 12390, pp. 56–84, 2020.
https://doi.org/10.1007/978-3-662-62308-4_3

their respective contents is also important. For instance, users of an online forum can share the same interests even without interacting often. They may be considered a community, but a structure based algorithm will not reach that outcome. On the other hand, a purely content based approach would exclude groups of users that interact often with each other if the contents of these interactions are varied, although logically they constitute a robust community.

We argue that the best approach for community detection is one that leverages both the structure and the content of a network to regulate community detection. Furthermore, to cope with the dynamic nature of social networks, modeling the problem as static is not appropriate and incremental approaches that also consider the past states of the network are required.

To this end, based on TILES [18], an incremental structure-based algorithm for fast overlapping community detection, we propose *COTILES*, that in addition to structure, also exploits network content. Our goal is to form more cohesive communities that still consist of groups of densely connected nodes but are also centered around the same sets of topics. The proposed algorithm considers a temporal labeled graph and defines the community label set that describes a community's contents. Besides the structural criteria TILES applies to determine when nodes form a community, COTILES applies a content-based constraint to ensure that the interaction that causes a node to be added to a community is relevant to the content of this community. In TILES, edges decay over time. Similarly, in COTILES, labels of edges and labels in community label sets also decay and expire with time to better capture the evolution of the topics (contents) of a community.

Furthermore, as each social network exhibits different characteristics, we design configurable variations of the basic COTILES approach that can be adjusted to best fit a particular application. In specific, we explore how tightening or relaxing the content-based or structure-based constraints alternatively can influence community detection. Therefore, we propose *Strict COTILES* that applies a more restrictive content-based constraint using a similarity threshold to tune it appropriately, and *COTILES with Enhanced Content* that relaxes the structural constraint enforced by COTILES. Both variations favor content against structure, aiming to derive more thematically cohesive communities. We also combine the two in a *Hybrid* approach. We compare the proposed algorithm COTILES against the original TILES using a real world dataset and show both quantitatively and qualitatively that COTILES leads to more cohesive clusters in contrast with TILES, in which community topics are far more scattered. Furthermore, we also explore the behavior of the three variations to determine advantages and weaknesses under different conditions.

The rest of the paper is structured as follows. In Sect. 2, we define basic concepts required for modeling the community detection problem under our setting and briefly describe the original TILES algorithm. In Sect. 3, we introduce COTILES and its three variations, while Sect. 4 includes our experimental results. In Sect. 5, we present related work distinguishing between structure-based, content-based and combined methods, while also presenting alternative categorizations one can consider. We conclude in Sect. 6, with a summary and directions for future work.

2   Preliminaries

We next define the basic concepts required for modeling our problem and describe the TILES [18] algorithm.

2.1   Basic Concepts

A social graph can be modeled as an undirected graph, $G = (V, E)$. Each node $u \in V$ of the graph corresponds to a user of the social network. Between these nodes there are edges $(u, v) \in E$ that correspond to interactions between users of the network. To model time evolution, edges in the graph require a timestamp $t_{(u,v)}$ that represents the time point the interaction occurs. Furthermore, to incorporate content, graphs are also labeled, so that for each edge (u, v), we have a label set $L_{(u,v)}$ that corresponds to the content of the interaction, which usually is a set of tags or text exchanged between nodes u and v. Thus, based on the definition in [2], we define:

Definition 1 (Temporal - Labeled Graph). *A temporal - labeled graph on G is an ordered quadruplet $G(T) = (V, E, L, T)$, where $T = \{T_e \subseteq \mathbb{N} : e \in E\}$ refers to the set of timestamps of the edges of G, and L refers to the global set of labels that appear in G at any time.*

The main task of graph analysis is the detection and monitoring of communities. We adopt a general, widely used definition of communities [14].

Definition 2 (Community). *A community is a group of nodes within which the connections are denser than those to other groups.*

2.2   The TILES Algorithm

Traditional community detection techniques can be used to monitor a graph's community structure through time. This is, in most cases, done by dividing the graph in snapshots based on - context dependent- temporal boundaries, detecting groups on each snapshot and then determining each community's continuation through consecutive snapshots.

Evolutionary community detection, on the other hand, aims to identify and maintain an up-to-date community structure of the graph throughout its evolution. TILES [18] is a fast evolutionary community detection algorithm that uses role propagation to distribute community membership to nodes based on their neighborhoods. Another advantage of TILES, is its capacity of allowing overlapping communities. Such as in real social relations, a person (node) is often member of more than one communities based on different criteria.

TILES identifies two roles of nodes participating in communities.

Definition 3. *A node is characterized as core if it is involved in at least one triangle with other nodes in the same community.*

That is, a core node belongs in at least one set with two other nodes of the same community, such that each node has a relationship to all other nodes in the set.

Definition 4. *A node is characterized as peripheral if it is an one-hop neighbor of a core node.*

A community may consist of both of these node categories. Core nodes are the main community representatives that spread community membership to their neighbors during role propagation. In contrast, peripheral nodes do not propagate their community membership to their own neighbors.

Edges connecting two nodes indicate interactions between them and they are characterized by a time-to-live (*ttl*) value. The default *ttl* value for an edge is set to infinite and it indicates edges that never expire such as friendships in a social network or co-authorship of an article in a collaboration network, while small *ttl* values indicate the fast decay of relations, hence communities, too. The latter are appropriate to model transient interactions such as exchange of messages in a communication network. In addition, the removal of older edges from the graph prevents memory growth problems.

The algorithm takes four parameters as input. The graph G which is initially empty, an edge streaming source S producing edges that correspond to interactions between the entities in the network, τ corresponding to the temporal observation threshold based on which the network evolution is monitored and the *ttl*. In every step of the algorithm based on a new edge (u, v) produced by S, there are four possible scenarios:

1. When both nodes u and v appear for the first time - or reappear after their expiration, no other action is performed.
2. When at least one node is a peripheral community node, that is a node which does not propagate community membership, no other action is performed.
3. When one node is a core node and the other appears for the first time in the graph, the second inherits a peripheral community membership from the first.
4. When both nodes are community cores, if they have common neighbors, their community memberships are re-computed, otherwise they propagate a peripheral community membership to each other.

The output of the algorithm is the community formation corresponding to each of the observation windows in the time period the network is monitored.

3   COTILES for Community Detection

We argue that relying only on structure or only on content for community detection ignores the other important parameter and thus may miss important analysis results. To this end, we propose COTILES, a community detection algorithm

---

Algorithm 1. $UpdateEdgeLabelsTimestamps(L_{new}, L_{curr})$

Require: L_{new} : New edge labelset, L_{curr} : Existing edge labelset
1: **for** $l \in L_{new} - L_{curr}$ **do**
2:      $l \cup L_{curr}$
3: **for** $l \in L_{new} \cap L_{curr}$ **do**
4:      $t_k = GetTimestamp(l, L_{new})$
5:      $SetTimestamp(l, L_{curr}, t_k)$

---

that combines both structure and content when determining communities in an evolving graph. We select TILES [18], an efficient incremental structural clustering algorithm and extend that to COTILES. COTILES, in addition to the structural constraints TILES enforces to form a community, also considers content-based constraints to form communities centered around sets of topics. Beyond that, COTILES maintains all major characteristics of TILES: it is incremental, it considers interactions expiration and supports overlapping communities. We also explore variations of the proposed COTILES algorithm by demanding greater content similarity among community members and relaxing the structural constraints. All proposed variations aim to derive communities with more cohesive content.

The output of COTILES and its variations follows basic TILES, presenting the community formation at each time slice corresponding to the observation window, with additional semantic information for each community, determined by the nodes inserted in it that are active at that time.

Next, we first describe the major contributions of COTILES for content management, and then present the overall extended algorithm. We conclude the section by presenting the proposed variations and how they differ from COTILES.

3.1   Label Management in COTILES

A social network in COTILES is modeled as a temporal labeled undirected graph, in which each edge has an accompanying label set describing its contents, and a timestamp denoting the time the interaction the edge represents occurred. The basic characteristic of COTILES is that it exploits these interaction labels, in addition to the structural information that the original TILES exploits, so as to form more thematically cohesive communities.

To describe the contents of a community, we first introduce the notion of the *Community Label Set* that is instrumental in the design of COTILES.

Definition 5 (Community Label Set). *Given a community c at time point t, L_c represents the set of labels that at that time describe the contents of c.*

As the community label set should represent the contents of the community, it is formed by considering the labels that correspond to the edges between the nodes that form the community.

Algorithm 2. $AddToCommunityLabels(C_o, L)$

Require: C_o : a community, L : a labelset
1: $L_{C_o} = GetLabels(C_o)$
2: **for** $l \in L$ **do**
3:      $t_k = GetTimestamp(l, L)$ ▷ Retrieve l's timestamp in L
4:      **if** $l \notin L_{C_o}$ **then**
5:          $L_{C_o}.Q.Add(l, t_k)$ ▷ insert new l with timestamp t_k
6:          $L_{C_o} = L_{C_o} \cup l$
7:      **else**
8:          $t_l = GetTimestamp(l, C_o)$ ▷ Retrieve l's timestamp in C_o
9:          **if** $t_l < t_k$ **then**
10:              $L_{C_o}.Q.Update(l, t_k)$ ▷ update timestamp of l to t_k

---

Algorithm 3. $LabelConstraint(u, v, C_o, G, clt)$

Require: u, v : nodes of graph G, $L_{(u,v)}$: labels of edge (u, v), C_o : an existing community, clt : community label set threshold.
1: $L_{C_o} = L_{comm}.GetLabels(C_o)$
2: **if** $L_{C_o} \cap L_{(u,v)} \neq \emptyset$ **or** $|L_{C_o}| \leq clt$ **then**
3:      $L_{comm}.AddToCommunityLabels(C_o, L_{(u,v)})$
4:      **return** true
5: **return** false

Lemma 1 (Community Label Set Construction). *Given a community c, consisting of a set of nodes N_c that form a set of edges E_c between them, its community label set, L_c is defined as: $L_c = \bigcup L_e, \forall e \in E_c$, where L_e is the set of labels attached to edge e.*

In COTILES, we assume that content and structure in a network are transient. User interests and their interactions and relationships, represented by network content and structure respectively, tend to decay over time. To model this decay, COTILES uses the time-to-live parameter for both labels and edges. In TILES, the ttl parameter is used only for edges so that edges expire after some time. In COTILES, an edge's labels inherit the edge's timestamp and the ttl parameter determines their lifetime as well. Furthermore, COTILES applies the same idea to model the lifetime of the labels that belong in a community label set. By using the time-to-live property not only for the edges, but for the labels in the community label set as well, we ensure that the communities evolve and do not remain restricted to the first topics and nodes that enter them.

In COTILES, the timestamps of the labels need to be updated similarly to how edge timestamps are updated if an edge $(e_{(u,v)})$ reappears before its expiration in TILES. However, when handling labels, as in COTILES, an additional concern is if the new label set of the edge, L_{new} shares labels with its current label set, L_{curr}. Thus, any label $l \in L_{new} - L_{curr}$ is added to the label set of $e_{(u,v)}$, and any label $l \in L_{new} \cap L_{curr}$ updates its timestamp, while the rest of the labels in L_{curr} maintain their timestamp. Thus, some labels in the set of an edge may expire before the edge itself. This procedure is detailed in Algorithm 1.

As it is evident, a respective update procedure should be deployed to handle a community's label set. Each time a new edge is added to a community, besides adding its labels in the community label set according to Lemma 1, there is the possibility that it carries labels already present in this community. Thus, the corresponding common labels update their timestamp in the community label set to the most recent value. Algorithm 2 details the process of updating the community label set. Given a community C_o with label set L_{C_o} and a set of labels L that corresponds to the edge that is incorporated in the community: if label $l \in L - L_{C_o}$, l is added to L_{C_o} with its corresponding timestamp t_k (lines 4–6), otherwise, l's timestamp is individually updated to the most recent value between its value in L and L_{C_o} (lines 9–10).

Besides describing the evolution of the contents of the communities through the definition of the community label set, the main contribution of COTILES is that, in addition to the structural criteria deployed by TILES to form a community, it also enforces content-based criteria via the *Label Constraint*. Our goal is to ensure that, when a node is added in a community, the interaction (edge), that causes this addition, has content similar to the content of the community. Content similarity can be defined using various methods and measures that aim to better capture semantic similarity. The similarity measure is orthogonal to our algorithm, therefore, different alternatives can be explored depending on their suitability for a given application domain. Here, we simply consider the intersection of two label sets to evaluate their similarity, i.e., considering two label sets equal if they share common labels. Consequently, before adding a node to a new community, we check whether the label set of the given edge has common labels with the community's label set.

To initialize the community label sets, as they start as empty, we use a community label set threshold, *clt*. If a community has fewer or equal to *clt* labels, then the Label Constraint is evaluated to true without actually computing the intersection of the community label set with the edge's label set. The purpose of this threshold is to encourage the formation of communities at the beginning of the algorithm, while also maintaining community cohesion and its value is context dependent.

The steps for the application of the Label Constraint are detailed in Algorithm 3. Given a community C_o, its labels are retrieved (line 1). The label set of the edge $L_{(u,v)}$ is checked against the label set of this community, L_{C_o} (line 2). If the intersection of the two sets is not empty or $|L_{C_o}| \leq clt$, the Label Constraint is evaluated to true and L_{C_o} is updated by Algorithm 2, otherwise the constraint is evaluated to false.

3.2   COTILES Description

We proceed with the overall COTILES algorithm, focusing on how we incorporate the novel content management processes, which we introduced in the previous section (Algorithm 1 to Algorithm 3), in the original TILES so as to make it content-aware.

Algorithm 4. COTILES.

Require: G : undirected graph, S : streaming source, τ : temporal observation threshold, ttl : edges & labels' time to leave, L : set of available labels, clt : community label set threshold.

1: $actual_t = 0$, $LRQ = \{\}$
2: **while** $S.isActive()$ **do**
3: $\quad (u, v, L_{u,v}, t_k) \leftarrow S.getNewInteraction()$
4: $\quad G.UpdateTimestamps(u, v, L_{e_{(u,v)}}, t_k)$
5: $\quad G.UpdateEdgeLabelsTimestamps(L_{e_{(u,v)}}, L_{LRQ.GetEdge(e_{(u,v)})})$
6: $\quad G.RemoveExpiredEdges(LRQ, ttl, actual_t)$ ▷ Remove expired edges and attached labels.
7: $\quad L_{comm}.RemoveExpiredLabels(ttl, actual_t)$ ▷ Remove expired labels from communities.
8: $\quad$ **if** $(u, v) \notin G$ **then**
9: $\qquad G.addEdge(e_{(u,v)})$
10: $\quad$ **if** $|\Gamma(u)| == 1$ **and** $|\Gamma(v)| == 1$ **then**
11: $\qquad$ **Continue**
12: $\quad core_u = G.GetCommunityCore(u)$
13: $\quad core_v = G.GetCommunityCore(v)$
14: $\quad$ **if** $core_u == \emptyset$ **and** $core_v == \emptyset$ **then**
15: $\qquad$ **Continue**
16: $\quad$ **if** $|\Gamma(u)| > 1$ **and** $|\Gamma(v)| == 1$ **then**
17: $\qquad G.PeripheralPropagation(u, \{v\}, clt, t_k)$
18: $\quad$ **else if** $|\Gamma(u)| == 1$ **and** $|\Gamma(v)| > 1$ **then**
19: $\qquad G.PeripheralPropagation(v, \{u\}, clt, t_k)$
20: $\quad$ **else**
21: $\qquad$ CN$=\Gamma(u) \bigcap \Gamma(v)$ ▷ CN: Common Neighbors
22: $\qquad$ **if** $|CN| == 0$ **then**
23: $\qquad\quad G.PeripheralPropagation(u, \{v\}, clt, t_k)$
24: $\qquad\quad G.PeripheralPropagation(v, \{u\}, clt, t_k)$
25: $\qquad$ **else**
26: $\qquad\quad G.CorePropagation(u, v, CN, clt, t_k)$
27: $\quad$ **if** $t - actual_t == \tau$ **then**
28: $\qquad OutputCommunities(G)$
29: $\qquad t = actual_t$

Algorithm 4 presents COTILES and includes the Label Constraint through the *Peripheral Propagation* mechanism. COTILES considers an undirected graph G, a streaming source of interactions S, a temporal observation threshold τ, a time-to-live threshold ttl, applicable both to edges and labels, a set of labels L that may appear, and finally the label community threshold clt.

To facilitate the edge expiration process in COTILES, similarly to the original TILES, we use a priority queue LRQ that, besides the edges, it also includes their labels (line 1). Method *UpdateTimestamps* (Algorithm 5) and *RemoveExpiredEdges* (Algorithm 6) handle the lifetime of the edges in TILES. In COTILES, in addition to these two procedures, we also include procedure

---

Algorithm 5. $UpdateTimestamps(u, v, L_{(u,v)}, t_k)$

Require: u, v : nodes forming edge uv, $L_{(u,v)}$: labels of edge $e_{(u,v)}$, t_k : timestamp of
    edge $e_{(u,v)}$ and its labels.
1: **if** $e_{uv} \in LRQ$ **then**
2:    $LRQ.Update(u, v, L_{(u,v)}, t_k)$ // update t_k ▷ Edge exists: Update timestamps
3: **else**
4:    $LRQ.Add(u, v, L_{(u,v)}, t_k)$

$UpdateEdgeLabelsTimestamps$ (Algorithm 1) and $RemoveExpiredLabels$ to handle the lifetime of the labels. The latter simply removes expired labels from the edge or community label sets they belong to.

Proceeding on the main functionality of COTILES, when the streaming source produces edge $e_{(u,v)}$ at t_k with label set $L_{(u,v)}$, there are four cases.

1. Both nodes u and v appear for the first time in the graph (lines 10–11). No other actions are performed until the next interaction is produced by the source S. This case is also described in TILES [18].
2. One node appears for the first time and the other is already existing but peripheral or both nodes are existing but peripheral (lines 12–15), in any case they do not belong to any community core. Since peripheral nodes are not allowed to propagate the community membership, no action is performed (none of the "if" clauses is satisfied) until the next interaction is produced by the source S. This case also applies to TILES.
3. Let u be a core node of a community and v appears for the first time (or vice versa) (lines 16–19). In this case, peripheral propagation is applied, as detailed in Algorithm 7, with the application of the COTILES Label Constraint.
4. Both nodes u and v are existing core nodes in G (lines 21–26).
   (a) Nodes u and v do not have common neighbors (lines 22–24). In this case, peripheral propagation takes place, as in the previous case with the application of the COTILES Label Constraint.
   (b) Nodes u and v have common neighbors (lines 25–26). In this case, core propagation that applies a variation of the Label Constraint, as described in Algorithm 8, takes place.

Peripheral propagation (Algorithm 7) before adding a node to the periphery of a community, first checks the COTILES Label Constraint (Algorithm 3) and only proceeds if that is evaluated to true.

Finally, core propagation in COTILES is described in (Algorithm 8). We will use the $I(\cdot)$ operator to represent the intersection of the communities of two, or more, nodes, and the $\Gamma(\cdot)$ operator to indicate the neighborhood of a node, i.e., the set of a node's one-hop neighbors. Core propagation, assumes that both u and v have at least one common neighbor z. If two nodes are core for a community, the third one becomes core (lines 10–21) as well and, the labels of the edges formed by the newcomer and the existing nodes are added to the community (lines 12, 16, 20) as long as the Label Constraint is satisfied, as ensured by the

---

Algorithm 6. *RemoveExpiredEdges(RQ, ttl, actual_t)*

Require: *LRQ* : a priority queue containing the edge candidate and its labels to be removed, *ttl* : edges time to live, *actual_t* : actual timestamp.

1: **for** $(u, v, t) \in RQ$ **do**
2:     **if** $(actual_t - t) \le ttl$ **then**
3:         $G.removeEdge(u, v)$
4:         $LRQ.remove(u, v)$ ▷ Also removes edge labels
5:         $to\_update = \{\Gamma(u) \cap \Gamma(v)\} \cup \{u, v\}$
6:         **for** $community \in I(u, v)$ **do**
7:             $components = G.getComponents(community)$
8:             **if** $|components| == 1$ **then**
9:                 $G.UpdateNodeRoles(community, to\_update)$
10:            **else**
11:                 **for** $c \in components$ **do**
12:                     $sc = G.NewCommunity(c)$
13:                     $G.RemoveNodes(community, V_c)$
14:                     $G.UpdateNodeRoles(sc, c)$

Algorithm 7. *PeripheralPropagation(u, nodes, clt, t_k)*

Require: *u* : node of G, *nodes* : a set of nodes, *clt* : threshold for Label Constraint, t_k : timestamp.

1: **for** $v \in nodes$ **do**
2:     **for** $c \in G.GetCommunityCore(u)$ **do**   ▷ ∀ community core nodes
3:         **for** $C_o \in c.GetCommunities()$ **do**
4:             **if** $LabelConstraint(u, v, C_o, G, clt) == true$ **then**
5:                 $G.AddToCommunityPeriphery(v, C_o)$

PeripheralPropagation method. Otherwise, we need to form a new community according to the original TILES. However, COTILES first applies a variation of the Label Constraint between the edges connecting the three nodes to ensure that they are content-wise related. In particular, if the intersection of the label sets of the edges that are connecting them is not empty, or their union has fewer or equal to *clt* labels, a new community, c^*, is formed (lines 2–9) and the labels of this community, L_{c^*} result from the union of the labels of the edges of the participating nodes (lines 4–5).

3.3 Tuning COTILES

The results of any community detection algorithm depend on the application domain and the nature of interactions and properties of the entities in the target social graph. The variety of approaches in the bibliography clearly shows that there is no unique solution appropriate for all applications. Our goal is to make COTILES tunable and adaptive to different requirements on community detection.

In addition to the time-to-live parameter, *ttl*, which determines the rate of time decay of edges and labels (i.e., interactions and content) and is also

---

Algorithm 8. $CorePropagation(u, nodes, clt, t_k)$

Require: u, v : node of G, CN : u&v common neighbors in G, clt : label threshold, t_k : timestamp.

```
1: for z ∈ CN do
2:     if I(u,v) == ∅ and I(u,z) == ∅ and I(v,z) == ∅ then
3:         if (L_(u,v) ∩ L_(u,z) ∩ L_(v,z)) ≠ ∅ or |L_(u,v) ∪ L_(v,z) ∪ L_(u,z)| ≤ clt then
4:             c* = G.CreateNewCommunity(u,v,z)              ▷ c*: new community
5:             L_c* = L_(u,v) ∪ L_(v,z) ∪ L_(u,z)            ▷ L_c*: labels of new community
6:             L_comm.Add(L_c*)
7:             PeripheralPropagation(u, Γ(u), clt, t_k)
8:             PeripheralPropagation(v, Γ(v), clt, t_k)
9:             PeripheralPropagation(z, Γ(z), clt, t_k)
```

```
10:     else if I(u,v) ≠ ∅ then
11:         G.AddToCommunityCore(z, I(u,v))
12:         L_comm.AddToCommunityLabels(C_z, L_I(u,v))
13:         PeripheralPropagation(z, Γ(z), clt, t_k)
14:     else if I(u,z) ≠ ∅ then
15:         G.AddToCommunityCore(v, I(u,z))
16:         L_comm.AddToCommunityLabels(C_v, L_I(u,z))
17:         PeripheralPropagation(v, Γ(v), clt, t_k)
18:     else if I(z,v) ≠ ∅ then
19:         G.AddToCommunityCore(u, I(z,v))
20:         L_comm.AddToCommunityLabels(C_u, L_I(z,v))
21:         PeripheralPropagation(u, Γ(u), clt, t_k)
```

---

used in TILES, COTILES introduces the community label set threshold, clt. This threshold, that is used for a community's label set initialization, enables nodes with no content similarity to form the initial communities based solely on their structural properties. As the algorithm proceeds with the addition of new interactions and the use of the ttl that removes old labels and edges from a community, the derived communities will be thematically cohesive, as only active topics and new interactions with contents around these active topics determine the evolution of the community.

Increasing the value of clt encourages the formation of larger communities at the start of COTILES that may include more than one thematic topics, while reducing its value leads to the formation of numerous smaller communities. Therefore, greater values are encouraged for networks where the content is more similar, while smaller values are most appropriate for networks with more diverse contents.

Besides the user-defined parameters that can be tuned according to the application requirements, as COTILES enforces both content and structural constraints for the formation of communities, we can explore variations that relax or tighten the aforementioned constraints to provide adjustable approaches appropriate for various settings.

Strict COTILES. In an effort to derive communities with greater content purity, we introduce *Strict COTILES* that restricts the insertion of nodes into communities in order to form groups with greater content similarity between their members. A possible weakness of COTILES, that Strict COTILES aspires to overcome, is grouping nodes to the same community only because they have few common labels with general meanings. This could happen due to the existence of some wide-spread labels in the network, connecting nodes not much semantically related, especially in social network applications that are centered around a particular domain of interest, i.e., all discussions in a technical support forum will be around technical issues.

Thus, Strict COTILES defines a more strict Label Constraint compared to COTILES. While the Label Constraint in COTILES allows a node to join a community c when the edge triggering the addition shares at least one label with the community label set of c, i.e., their intersection is not empty, Strict COTILES requires that the cardinality of the intersection of the two labels sets is equal or above a user defined threshold ($StrT$). By restricting the Label Constraint, higher similarity is achieved within a community. Strict COTILES uses the same mechanism for encouraging the initial formation of communities by utilizing *clt* similarly to COTILES. The *Strict Label Constraint* algorithm is illustrated in Algorithm 9.

This Strict COTILES variation follows the original COTILES procedure only deploying the Strict Label Constraint instead of the Label Constraint in peripheral propagation (Algorithm 7, line 4). Furthermore, it correspondingly alters the variation of the Label Constraint that is applied in core propagation (Algorithm 8, line 3). Similarly to the Strict Label Constraint, the cardinality of the intersection of the label sets of the edges connecting the three nodes that are candidates for forming a new community is required to be greater or equal to the $StrT$ threshold.

While the use of the $StrT$ threshold introduces a new parameter in the algorithm, it allows us to tune Strict COTILES according to the application requirements. COTILES offers one extreme in the spectrum where $StrT = 1$ and at least one shared label is enough to allow an edge (u, v) to be incorporated in a community. On the other hand, setting $StrT$ equal to $|L_{(u,v)}|$, that is requiring that the label set of the new edge is a subset of the community label set determines the other extreme, requiring the contents of the edge to already be included in the contents of a community to incorporate it.

The Strict Label Constraint can be tuned to cover cases between COTILES and Strict COTILES with maximum strict threshold. In some cases, COTILES tends to add nodes into communities with barely similar content caused by only one general common label, as we mentioned, whereas Strict COTILES with maximum strict threshold has an absolute similarity criterion and excludes many nodes from a community where they belong in terms of content. Strict COTILES with the appropriate context-related threshold is quite more selective than the main algorithm, but does not require all of the edge's labels to be already present inside the community label set. As an appropriate threshold we propose using an estimation of the average of the similarity values within the label sets of the edges of a community.

Algorithm 9. $StrictLabelConstraint(u, v, C_o, G, clt, StrT)$

Require: u, v : nodes of graph G, $L_{(u,v)}$: labels of edge (u, v), C_o : an existing
   community, clt : community label set threshold, $StrT$: Strict threshold.
   1: $L_{C_o} = L_{comm}.GetLabels(C_o)$
   2: **if** $L_{C_o} \cap L_{(u,v)} \geq StrT$ **or** $|L_{C_o}| \leq clt$ **then**
   3:    $L_{comm}.AddToCommunityLabels(C_o, L_{(u,v)})$
   4:    **return** true
   5: **return** false

COTILES with Enhanced Content. *COTILES with Enhanced Content, EC,*
is another variation that also aims to derive communities with more similar
contents. However, instead of restricting the content based constraint to achieve
this, in EC, we relax one of the structural constraints COTILES applies. The
main idea behind this variation is that in some cases detecting nodes with similar
content is more important than relying on the density of the interactions among
them. For instance, consider a case where many users share the same interests
but while they interact together, they mostly interact within subgroups. It might
be of value to form a single larger community in such a case.

EC intervenes in the main procedure of COTILES and shifts the attention
mostly at content-based relations between nodes. The Label Constraint is applied
as in COTILES to ensure content similarity. The difference is that EC allows
peripheral nodes to propagate community membership to their neighbors, as long
as the Label Constraint is satisfied. Thus, EC applies the main COTILES algorithm as described, in Algorithm 4, but line 12 to line 15, which check whether
one of the nodes of the new edge are core, are omitted and the algorithm continues with peripheral and core propagation. Thus, even if the two nodes of the
new edge are both peripheral, they can propagate community membership to
one another if the new edges' labels are similar to any of the communities they
have joined.

We expect EC to encourage the formation of larger communities that include
nodes that may not be so densely connected but share similar contents.

Hybrid COTILES. The final variation we consider is a hybrid approach that
combines the two newly proposed variations, namely Strict COTILES and EC,
in order to exploit and combine the advantages of each one.

Strict COTILES mandates a higher similarity between two label sets to consider them similar by deploying a user defined threshold, while EC relaxes the
structural constraints that COTILES applies allowing not only core, but peripheral nodes as well to propagate community membership. In the hybrid approach,
we combine the two, thus both relaxing the structural constraints as in EC, but
also demanding higher content similarity in a community's members as in Strict
COTILES.

To apply the hybrid approach, COTILES is altered as follows. First, the
change that EC applies in COTILES by omitting the check of whether any of

the two nodes of a new incoming edge are core, is also applied in the hybrid approach, that is, lines 12 to 15 of Algorithm 4 are omitted. Thus, even if the nodes are peripheral, they can propagate community membership to one another as long as the new edge's label set is similar to any of the nodes' community label sets. However, to determine their similarity, instead of the Label Constraint, the Strict Label Constraint as described in Algorithm 9 of the Strict COTILES approach is applied in peripheral propagation. Similarly, in core propagation the variation of the Label Constraint that is described in Strict COTILES is also deployed in the hybrid approach, ensuring both times that the cardinality of the intersection of the compared label sets is greater or equal to the $StrT$ threshold.

Compared to Strict COTILES, the Hybrid approach will allow communities with less dense interactions among their members due to EC that relaxes structural constraints. At the same time, it will ensure higher content purity within them compared to EC, due to Strict COTILES that ensures higher content similarity between community members.

4   Evaluation

We compare COTILES and its variations against the original TILES measuring both structural and content-based properties of the derived communities, focusing, also, on how these properties change, as the communities evolve through time.

In order to make content characteristics countable for TILES as well, we apply COTILES without the label constraints, deriving the communities formed by the original TILES but associated with their corresponding community label sets.

Measuring different structural properties, such as community cardinality and density, we select to report our results on cardinality as they showcase clearly the differences between the different variations of our approach. We also separately study the overlap between the formed communities, as this is one of the most interesting dimensions of TILES and we are interested in exploring the influence of content awareness, as introduced by COTILES, on the derived results. As far as the content-based properties are concerned, we measure the number of unique labels in each community and also content similarity between communities. Both these measures demonstrate how well we succeed with respect to our primary goal, that is, to derive cohesive well-separated with regards to content communities.

4.1   Experimental Setup

We use a dataset retrieved from the *Unix.StackExchange.com* forum available via the *Stack Exchange Network* [12] and construct a temporal-labeled graph, $G(T) = (V, E, L, T)$, where V corresponds to forum users, E to interactions between users occurring when members answer or comment to each other's posts,

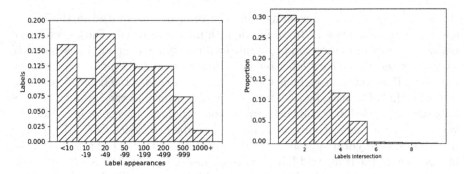

Fig. 1. (**Left**) label distribution, and (**right**) similarity between corresponding edge and community label sets.

L to the set of labels used by the forum's users as tags describing their posts, and T to the set of timestamps for the interactions.

The constructed graph has 542120 edges formed between 87438 nodes. 2615 different labels appear through the graph's lifetime, with a total amount of 1533354 occurrences. The distribution of labels is illustrated in Fig. 1 (left). The observation timeline lasts 6 years, from August 2010 to August 2015.

Threshold and Setup Selection. Before evaluating the proposed algorithms, we first need to determine the appropriate values of our parameters so as to detect communities in the forum-based network with best structural and content coherence.

The first threshold we examine is the Community Label Set threshold (clt), which is responsible for the initialization of the communities and their label sets, and expect it to be highly context dependent. Setting a high clt value in graphs with few labels per edge would lead to lower content purity in communities as nodes with different interests may join the community before the Label Constraint is enforced. On the other hand, setting clt too low might lead to very few communities. We experiment with different values of clt varying from 3 to 10, and measure the cardinality (Fig. 2(left)) and the content similarity (Fig. 2(right)) of the detected communities. Content similarity is measured using the Jaccard similarity of the community label sets for the communities that survived in consecutive snapshots. In Fig. 2 (right), the horizontal axis refers to the average Jaccard similarity of a community's label set between one time slice and its consecutive, if the community survives in it. The vertical axis refers to the percentage of communities reaching that similarity. A value of $clt = 5$ leads to higher content similarity as the proportion of instances with similarity near 1 is increased according to Fig. 2(right), and the ones with similarity near 0 are reduced compared to other clt values. Considering Fig. 2(left), that presents the histogram illustrating the cardinality distribution among the communities, we can observe again that a value of $clt = 5$ deters the formation of highly extended

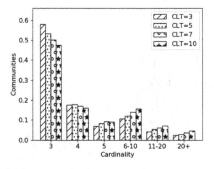 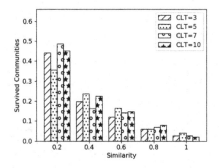

Fig. 2. Varying community label set threshold: (**left**) community cardinality, and (**right**), content similarity in time.

communities (20+ members). Therefore, for the rest of our experiments, we set *clt* to 5.

Next, we appropriately choose an observation window (τ) and time-to-live (*ttl*) value. In Fig. 3, we illustrate the fluctuation of average community cardinality in time with three different setups with different values for τ and *ttl* for COTILES. SETUP 1 corresponds to *ttl* = 15 and a corresponding τ = 30, i.e., double the size of *ttl*. SETUP 2 has equal value for *ttl* and τ (= 30). Finally, SETUP 3 has a *ttl* = 30 and a larger τ = 45. Taking into account the need for fast execution and less information loss, we prefer SETUP 3 over SETUP 1 and 2 respectively. Our argument is also amplified by the plots' slopes, where although all 3 SETUPs achieve similar average community cardinality, SETUP 3 avoids the high fluctuations between time slices the other SETUPs exhibit. This SETUP allows edges and labels to expire if they are not renewed within the observation window as τ is greater than the selected *ttl*. Thus, it is appropriate for networks with transient relationships as the ones observed in a forum.

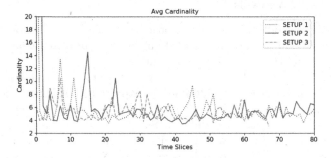

Fig. 3. Average community cardinality in time.

Finally, with regards to our variations, we also need to select an appropriate strict threshold value *StrT* for both Strict COTILES and the Hybrid approach.

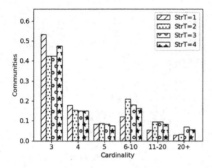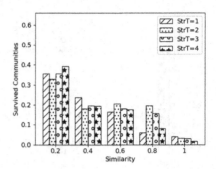

Fig. 4. Strict COTILES with different threshold: community cardinality (**left**), content similarity between same communities in time (**right**).

In Fig. 1(right), we count the overlap of the edges label sets and the label sets of the communities they enter according to COTILES. We observe that indeed, a big percentage of edges have low content similarity with the groups they enter and a restriction policy could be useful. The average recorded value corresponds to 2.39 labels in common, so a valid $StrT$ value would be 2 to 3. To acquire further evidence, we perform the same experiment we performed when tuning clt. Figure 4(left) corresponds to the effects of $StrT$ in community cardinality, while Fig. 4(right) corresponds to the content similarity of communities in consecutive windows. Note that again, the y-axis in both figures shows the percentage of communities reaching a specific level of cardinality and similarity, respectively. The experiments show that a value of 3 is not inferior to the other values and is also closer to the average labels per edge in our setting which is measured to 2.82. Thus, we select $StrT = 3$ so as to differentiate the variations significantly from COTILES to better study their effect. Note that a strict threshold equal to 1 corresponds to COTILES.

4.2   Structural Analysis

We now discuss the outcomes of our empirical evaluation focusing on the behavior of COTILES and its variations based on the structural characteristics of the formed graphs and the respective communities.

In the left segment of Table 1, we observe the average cardinality of communities throughout the network's lifetime. As expected, COTILES tends to form more communities than TILES per time-slice as the label constraint makes joining an existing community more difficult. Strict COTILES detects 59 communities on average in each time-slice, while EC quite less, with the hybrid method falling between them. We also include Strict MAX, which refers to the Strict variation with maximum strict threshold as requiring the exact same contents to join a community is an extreme case worth investigating. As expected, this leads to the highest number of detected communities. Taking into consideration the average cardinality of communities shown in the second column of Table 1, we observe the trend of forming more communities but with smaller cardinality.

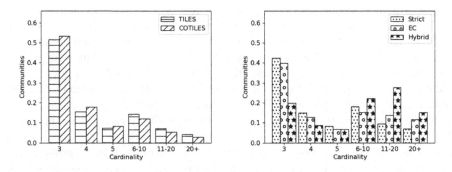

Fig. 5. Community cardinality: (**left**) TILES vs. COTILES, and (**right**) variations.

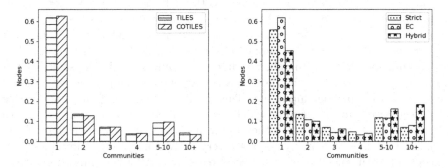

Fig. 6. Node community membership: (**left**) TILES vs. COTILES, and (**right**) variations.

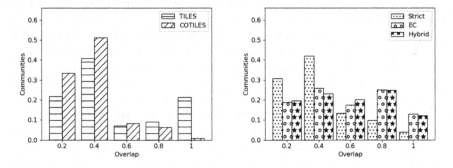

Fig. 7. Community overlap: (**left**) TILES vs. COTILES, and (**right**) variations.

Histograms in Fig. 5(left) and Fig. 5(right) illustrate the percentage of communities with the corresponding cardinality for COTILES compared to TILES and the three variations respectively. COTILES differentiates from the original TILES algorithm by detecting a higher percentage of small communities. Its variations follow the same trend, except for the hybrid algorithm, which leads to a high percentage of big communities. Note that the results are not strictly comparable, as the percentages refer to the total number of communities per algorithm that differ. For example, COTILES detects 5141 communities in the networks' lifetime with the corresponding setup, while the hybrid method only half (2560). Thus, the absolute number of big communities is the same and one can claim that this method is suitable in cases where analysis of groups with high cardinality is needed, with guaranteed robust content relation between their members.

An important advantage of TILES, compared to other community detection algorithms and a serious motivation for us to extend it, is its potential in discovering overlapping communities, letting nodes to be part of different groups. In our next set of experiments, we discuss the behavior of TILES with respect to this feature and how COTILES and its variations manage such situations.

Node community membership is illustrated in Fig. 6(left) and Fig. 6(right) and refers to the number of communities that each node is a member of. The x-axis corresponds to the number of different communities in the same slice in which a node is present while the y-axis corresponds to the percentage of nodes that reach this membership value. The specific dataset contains many nodes not assigned to any community which are not pictured, as many users tend to interact with others only when they are in need of answers. At the same time, usual in forums is that many users are very active, answering many questions. Except for their knowledge, they are also motivated by the importance their contribution may have, as high reputation in a forum from Stack Exchange is valuable. Similarly to the previous experiment, COTILES achieves to generally assign more nodes to only one or few communities and less nodes are highly distributed compared to TILES (Fig. 6(left)). COTILES variations and especially those with enhanced content (EC, hybrid) assign nodes to more communities, as their premise is to give a node a higher chance to enter a group with related labels (Fig. 6(right)).

Figure 7 considers a different perspective and measures community overlap. The horizontal axis represents the overlap as the percentage of common members between communities. Zero overlap represents total dissimilarity, while overlap equal to 1 would indicate totally overlapping communities. The vertical axis represents the percentage of communities where the respective amount of overlap appears. In general, COTILES (Fig. 7(left)) and its variations (Fig. 7(right)) succeed in reducing the percentage of communities with the highest overlap with respect to TILES. But the variations exhibit higher overlap values between 0.4 and 0.8, as we expect after viewing the results of the previous experiment measuring node community membership. Thus, one should consider when it is appropriate to encourage nodes to join communities with similar contents by also considering the tradeoff with respect to an increased overlap between the derived communities.

4.3   Content-Based Analysis

To validate our structural analysis results, we also directly investigate the content of the communities formed by all algorithms.

Starting from the third column of Table 1, where the average number of unique labels in each community is shown, it is clear that, when total content community purity is needed, Strict COTILES with maximum label constraint (Strict MAX) is the appropriate algorithm. We retrospect that, in this case, a new edge enables one node to enter a community only if all of the edge's labels are already present in the community label set. Also, COTILES succeeds in significantly reducing the size of community label sets, so that they are more focused. All three variations we examine with $StrT = 3$ when used tend to form wider label sets for their communities, which however is justified as they also form communities with higher cardinality. In addition, the quite high portion of labels in the label sets of communities formed by our hybrid method is a result of their high cardinality and also very high standard deviation (17.9965) contrary to TILES (8.0345) and COTILES (3.9946) for example.

Table 1. Algorithm variations statistics.

	#Coms	Avg. cardinality	Avg. #Labels
TILES	59.0588	6.0283	23.5955
COTILES	75.6029	5.2466	9.7947
Strict	59.1470	6.8660	21.1152
Strict MAX	92.1014	3.4689	3.6266
EC	29.2647	8.0022	20.8225
Hybrid	37.6470	10.124	46.3676

The plots in Fig. 8(left) and Fig. 8(right) show how similar in terms of content are the communities of respective time-slices. The horizontal axis corresponds to the average Jaccard similarity between a community's label set and others' in the same time period. Values in this axis correspond to upper thresholds and portions near zero correspond to totally different set of labels. The vertical axis corresponds to the percentage of communities achieving the respective amount of similarity between them and others on average. COTILES is rather better than TILES in distinguishing communities based on their content, as it has higher percentages of communities with low content similarity (<0.2). Strict COTILES follows the same behavior as shown in Fig. 8(right), while the two variations containing the enhanced content procedure show different results. Their higher cardinality and lower number of communities result to hyper-communities with reasonable similarity between their content. Furthermore, the higher overlap they exhibit also accentuates this effect.

Finally, in terms of content based criteria, histograms in Fig. 9(left) and Fig. 9(right) illustrate how the contents of a community are preserved through

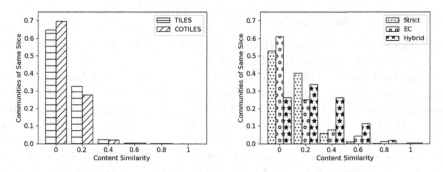

Fig. 8. Content similarity between communities: (**left**) TILES vs. COTILES, and (**right**) variations.

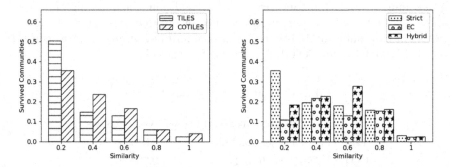

Fig. 9. Content similarity between the same community in time: (**left**) TILES vs. COTILES, and (**right**) variations.

time. The horizontal axis refers to the average Jaccard similarity of a community's label set between one time slice and its consecutive, if the community survives in it. The vertical axis refers to the percentage of communities reaching that similarity. COTILES and all three of its variations achieve better results compared to TILES. As shown in the first histogram (Fig. 9(left)), half of the communities detected by TILES have totally different content if they succeed in surviving to the next time-slice. On the other hand, communities in COTILES and its discussed variations are highly correlated in terms of content between consecutive time-slices (Fig. 9(right)). Our hybrid method which combines Strict COTILES with EC seems to be the most suitable for content preservation. The results are also not beautified by the conversion into percentages, since the hybrid method has 287 survived instances compared to 318 of COTILES and contrary to nearly the half (152) of TILES.

4.4 Qualitative Results

In order to further investigate the algorithms' behavior, we focus on sample communities and present comparisons between snapshots of communities detected by TILES, COTILES and its variations in the same time slice. We track the same communities based on their membership and collocate their community label sets. Mutual emergence of members and labels is noted by bold font in the following tables.

Table 2 compares a community detected by TILES and its corresponding community detected by COTILES. As expected, the community in the second case has lower cardinality and limited label set compared to the first case. Focusing on the two label sets, we observe that COTILES achieves higher content cohesion by containing a lower amount of labels, however with high number of appearances in most of them. In contrast the community label set of the corresponding community detected by TILES contains 9 labels with only one appearance, a fact underlining content fragmentation.

Table 2. Community detected by (left) TILES and (right) COTILES.

	TILES	COTILES
Members	4, 1002, 1131, **1207**, **1327**, **1571**, 2511, **2534**	**1207, 1327,** **1571, 2534**
Labels	'linux': 4, '**filesystems**': 4, '**mount**': 4, '**windows**': 3, '**virtual-machine**': 3, 'cron': 2, 'fedora': 2, 'gnome': 2, 'login': 2, 'gdm': 2, '**debugging**': 1, '**suspend**': 1, 'grep': 1, 'io-redirection': 1, 'ffmpeg': 1, 'system-installation': 1, 'grub2': 1, 'install': 1, 'package-management': 1	'**filesystems**': 4, '**mount**': 4, '**windows**': 3, '**virtual-machine**': 3, 'linux': 1, '**debugging**': 1, '**suspend**': 1

In the following three tables, we compare communities between COTILES and its variations to distinguish cases in which they can be selected.

Table 3 compares communities from COTILES and Strict COTILES respectively. Members and labels of the second community are subsets of the ones observed by the first community. In fact, Strict COTILES detects a subgroup

Table 3. Community detected by (left) COTILES and (right) Strict COTILES.

	COTILES	STRICT
Members	**16792, 22257, 38906**, 65304, 86440, 117549, **246185**	**16792, 22257, 38906, 246185**
Labels	'**bash**': 9, '**printf**': 3, '**command-substitution**': 2, '**subshell**': 2, '**quoting**': 2, 'process-substitution': 1, 'shell': 1, 'shell-script': 1, 'javascript': 1, 'base64': 1, 'tar': 1	'**bash**': 5, '**printf**': 3, '**quoting**': 2, '**subshell**': 2, '**command-substitution**': 1

inside the community of COTILES and is suitable for cases where higher resolution is needed.

Table 4 focuses on communities formed by COTILES contrary to its variation with Enhanced Content. The opportunity given to nodes in the EC variation to propagate community membership to their neighbors without necessarily being core nodes, leads to groups with higher cardinality but also aspires to preserve content coherence. Indeed, in this example, a new member is added to the community by EC, while its community label set is not scattered, exhibiting the advantage of this method when relaxed structural connections are tolerated in order to discover content similarities.

Table 4. Community detected by (left) COTILES and (right) COTILES EC.

	COTILES	EC
Members	**5280, 8979, 13003, 19294**	**5280, 8979,** 11318, **13003, 19294**
Labels	'debian': 3, 'mercurial': 2, 'certificates': 2, 'apt': 2, 'package-management': 2, 'software-installation': 1, "textbfgimp': 1	'debian': 6, 'software-installation': 3, 'gimp': 3, 'package-management': 2, 'apt': 2, 'mercurial': 2, 'certificates': 2

Table 5. Community detected by (left) COTILES and (right) Hybrid COTILES.

	COTILES	HYBRID
Members	**18887, 170373, 311452**	**18887,** 120884, **170373, 311452,** 316025
Labels	'regular-expression': 5	'linux': 6, 'tar': 3, 'process': 3, 'exec': 3, 'exit': 3, 'vfork': 3, 'scripting': 2, 'backup': 2, 'debian': 2, 'linux-kernel': 2, 'shared-library': 2, 'elf': 2, 'sudo': 1, 'bash': 1, 'shell-script': 1, 'timestamps': 1

Finally, a community detected by our hybrid method which combines enhanced content technique with the Strict Label Constraint is shown in the second column of Table 5, next to a corresponding community of COTILES. The enhanced content part leads to a bigger community by 2 members. The only label of the first community label set ('regular-expression') is not present in the second case, as the strict constraint requires at least 3 labels of incoming edges to match the community label set. This shows that by allowing more nodes to join the community, the hybrid method succeeded in determining a content shared by more nodes.

Concluding, we can see that under different considerations, each variation offers advantages and their application should depend on the characteristics of the underlying network. However, COTILES and all variations manage to derive communities with higher content similarity compared to the original TILES.

5 Related Work

We organize related research into three categories: structure based, content based and combined approaches for community detection. Then, we present recent surveys that provide different categorizations on related research on community detection in evolving graphs.

5.1 Structure-Based Methods

The most popular category, structure-based clustering algorithms, assesses the links between nodes as the main criterion for a community. The survey in [9] studies the problem for static networks in extent, while the one in [11] considers recent approaches in evolving graphs. In [24], community evolution is evaluated using metrics such as growth or disappearing rate. The work in [16] focuses on the life expectancy of communities with regard to the weight of their intra- or inter-community edges. Based on clique percolation, the authors focus in a scientists collaboration network and a mobile phone users network and conclude in correlations between communities' life expectancy and member composition. In [20], a snapshot-based approach is used to track the evolution of communities and interesting persistent communities with high cohesion are detected. In [15], a method that uncovers intrinsic communities, which are subsets of communities with higher connection and importance, is proposed. Di Tursi et al. [8] propose a filter-and-verify framework using a time-and-graph-aware locality sensitive hashing to efficiently discover promising community cores, and focus on varying densities of interactions, only considering, graph structure.

Evolutionary clustering is proposed in [5] based on the idea that clusters should be calculated not based on the information of a single snapshot, but considering also the previous states of the graph. TILES [18], we extend in this work, is also an example of evolutionary clustering. Another algorithm falling into this category is the LabelRankT algorithm introduced by Xie et al. [26]. In this case, labels refer to node identifiers, and do not represent content as in our case. A community is formed considering the nodes having the same label, edge weights and directions.

5.2 Content-Based Methods

As the content of the network is also important in order to identify cohesive groups, scientific research has also shown interest in this area, mostly, though, for static networks. In [3], time-sensitive tag clustering techniques for finding semantically similar tags are explored. Similarly, in [22], co-occurrence analysis and clustering techniques are used in two real tag sets to derive meaningful groups of tags and identifying relationships between subsets of such tags. In [23], communities are distinguished between structural and content-based ones and a non-overlapping method is proposed to identify user groups with similar interests. In [10], semantics and social features are exploited in addition to content analysis so as to produce tag clusters. In contrast to the previous static approaches, in [1], an online approach analyzes micro-blogging streams (from Twitter) by modeling the problem as discovering dense clusters in dynamic graphs and using the correlated keyword graph structure.

5.3   Combined Methods

Few methods, similarly to ours, take advantage of both structure and content for community detection, which can lead to a better understanding of communities' activities and their emerging trends. In [7], authors are motivated by the fact that scientific literature is basically clustered into journals and conferences with predefined domains. To this end, techniques to extract the main context of each paper are proposed. In [25], the network is transformed into a Node-Edge Interaction network which captures both the linkage structure as well as node and edge content. A random walks-based algorithm is then used to detect dynamic communities. As neither edges or nodes expire, this approach may not accurately model real world conditions, where interests in topics and relationships in a community may decay over time.

Sadri et al. [21] analyze tweets and user mentions, introducing three models for user interest inference. Among them, the community interest pattern model deals with both community structures and user interests. Nevertheless, the focus of this work is on identifying interests and not on community evolution per se. In [4], the authors introduce an interest social network model to connect nodes with similar interests. If two nodes are connected with an interest in the same topic, a new link is formed between them. Community discovery is then applied on this updated graph that represents both structural and content-based information. Again, this work focuses on community detection and does not address community evolution.

Finally, this paper extends and verifies the work in [19], where COTILES is first presented. Here, we first introduce the three variations of COTILES that are designed as more configurable methods that can be better tuned to suit the underlying network characteristics, while still relying on the core idea of COTILES of combining structural and content-based criteria. With our thorough experimental evaluation, we provide comparative results that verify the utility not only of COTILES compared to the original TILES, but also of all three of the proposed variations as well.

5.4   Categorizing Related Methods

There have been recent surveys examining the developments in our area of interest offering categorizations of the related works under a variety of criteria that influence how we deal with community evolution.

In [6], related studies on the problem of tracking community evolution over time in dynamic social networks are presented. This paper provides a categorization of existing methods into four classes, based on the approach each method follows for achieving community evolution tracking while describing strengths and weaknesses for each. The first category is comprised of methods based on independent successive static detection and matching. The second class includes methods using dependent successive static detection. The third class consists of methods that perform simultaneous study of all stages of community evolution. Finally, the fourth category is based on methods working directly on temporal networks.

Before that, in [17], methods for community discovery in dynamic networks are enumerated and grouped into three basic classes, namely, *Instant Optimal*, *Temporal Trade-off* and *Cross-Time* community detection, each of them comprising of several subclasses. The category of Instant Optimal community detection, considers that communities at time step t only depend on the current state of the network at t. Matching communities found at different steps might involve looking at communities found in previous steps, or considering all steps, but communities found at t are considered optimal with respect to the topology of the network at t. In the category of Temporal Trade-off community detection, communities defined at an instant t depend not only on the topology of the network at that time but also on the past instances of the topology, past partitions found, or both. Communities at t are therefore defined as a trade-off between an optimal solution at t and the known past. Regarding Cross-Time community detection, the authors consider methods that take into account the whole network evolution. Methods of this class search a single partition directly for all time steps. Communities found at t depend not only on the past like in the Temporal Trade-off category, but also on future instances and modifications that follow t in time.

In another very recent survey, Interdonato et al. [13] present a taxonomy of networks that are feature rich, or, in other words, graphs that feature some additional characteristics beyond their topology. In particular, they consider *Attributed Graphs*, where edges or nodes hold generic attributes, *Heterogeneous Information Networks*, that is, networks modeling heterogeneous node and edge types, *Multilayer Networks* representing different online/offline relations between the same set of users, *Temporal Networks* modeling discrete or continuous time aspects in networked data, *Location-aware Networks* holding location information, and finally, *Probabilistic Networks* modeling uncertain relations.

6   Conclusions

In this paper, we proposed COTILES, an online community detection algorithm that relies on both the structural and content-based information in a social network. COTILES extends TILES [18], a purely structure based community detection algorithm that allows the detection of overlapping communities with time decaying edges. COTILES maintains the advantages of TILES, but also aims to derive more thematically cohesive communities by applying a content based constraint that regulates community formation. We also presented configurable variations of the proposed algorithm that either relax or tighten the enforced structural and content constraints so as to be more appropriate for various social networks and application domains. Strict COTILES demands higher content purity in communities using an appropriate threshold, while Enhanced COTILES relaxes the structural constraints of COTILES favoring the detection of content-based communities. Finally, a hybrid approach combines the two variations. Through our experimental results, we showed that COTILES derives labeled communities that are more thematically cohesive compared to the original TILES. We also showed the advantages offered by each variation under different conditions.

In our current work, we focused on the quality of the derived results testing different configurations and proposing context-dependent variations. Next, we plan to focus on performance, with respect to the time and space complexity of our method. In particular, we will aim to minimize the overhead that content management imposes on COTILES compared to TILES. In this regard, we plan to parallelize COTILES and deploy it exploiting available big data parallel processing platforms so as to evaluate its performance. Our future plans also include exploring alternative ways to measure the similarity between the content of nodes and communities. We will move in two directions towards this end. Firstly, we are going to enrich the labels describing the contents of the network by applying appropriate NLP techniques and semantic processing based on the use of thesaurus. Finally, we are also going to evaluate the use of different similarity measures that better capture semantic similarity between sets of labels.

References

1. Agarwal, M.K., Ramamritham, K., Bhide, M.: Real time discovery of dense clusters in highly dynamic graphs: identifying real world events in highly dynamic environments. Proc. VLDB Endow. **5**(10), 980–991 (2012)
2. Akrida, E.C., Gąsieniec, L., Mertzios, G.B., Spirakis, P.G.: On temporally connected graphs of small cost. In: Sanità, L., Skutella, M. (eds.) WAOA 2015. LNCS, vol. 9499, pp. 84–96. Springer, Cham (2015). https://doi.org/10.1007/978-3-319-28684-6_8
3. Begelman, G., Keller, P., Smadja, F., et al.: Automated tag clustering: improving search and exploration in the tag space. In: Proceedings of the Collaborative Web Tagging Workshop at 2006 World Wide Web Conference, pp. 15–33 (2006)

4. Bu, Z., Zhang, C., Xia, Z., Wang, J.: A fast parallel modularity optimization algorithm (FPMQA) for community detection in online social network. Knowl.-Based Syst. **50**, 246–259 (2013)
5. Chakrabarti, D., Kumar, R., Tomkins, A.: Evolutionary clustering. In: Proceedings of the 12th ACM SIGKDD International Conference on Knowledge Discovery and Data Mining, pp. 554–560. ACM (2006)
6. Dakiche, N., Tayeb, F.B.S., Slimani, Y., Benatchba, K.: Tracking community evolution in social networks: a survey. Inf. Process. Manag. **56**(3), 1084–1102 (2019)
7. De Nart, D., Degl'Innocenti, D., Basaldella, M., Agosti, M., Tasso, C.: A content-based approach to social network analysis: a case study on research communities. In: Calvanese, D., De De Nart, D., Tasso, C. (eds.) IRCDL 2015. CCIS, vol. 612, pp. 142–154. Springer, Cham (2016). https://doi.org/10.1007/978-3-319-41938-1_15
8. Di Tursi, D.J., Ghosh, G., Bogdanov, P.: Local community detection in dynamic networks. In: Proceedings of the 2017 IEEE International Conference on Data Mining (ICDM), pp. 847–852. IEEE (2017)
9. Fortunato, S.: Community detection in graphs. Phys. Rep. **486**(3), 75–174 (2010)
10. Giannakidou, E., Kompatsiaris, I., Vakali, A.: SEMSOC: semantic, social and content-based clustering in multimedia collaborative tagging systems. In: Proceedings of the 2008 IEEE International Conference on Semantic Computing, pp. 128–135. IEEE (2008)
11. Hartmann, T., Kappes, A., Wagner, D.: Clustering evolving networks. In: Kliemann, L., Sanders, P. (eds.) Algorithm Engineering. LNCS, vol. 9220, pp. 280–329. Springer, Cham (2016). https://doi.org/10.1007/978-3-319-49487-6_9
12. Stack Exchange Inc.: Stack exchange data dump. https://archive.org/details/stackexchange. Accessed 10 Feb 2019
13. Interdonato, R., Atzmueller, M., Gaito, S., Kanawati, R., Largeron, C., Sala, A.: Feature-rich networks: going beyond complex network topologies. Appl. Netw. Sci. **4**(1), 4 (2019)
14. Jdidia, M.B., Robardet, C., Fleury, E.: Communities detection and analysis of their dynamics in collaborative networks. In: Proceedings of the 2nd International Conference on Digital Information Management, pp. 744–749. IEEE (2007)
15. Nath, K., Roy, S.: Detecting intrinsic communities in evolving networks. Soc. Netw. Anal. Min. **9**(1), 13 (2019)
16. Palla, G., Barabási, A.L., Vicsek, T.: Quantifying social group evolution. Nature **446**(7136), 664 (2007)
17. Rossetti, G., Cazabet, R.: Community discovery in dynamic networks: a survey. ACM Comput. Surv. (CSUR) **51**(2), 1–37 (2018)
18. Rossetti, G., Pappalardo, L., Pedreschi, D., Giannotti, F.: Tiles: an online algorithm for community discovery in dynamic social networks. Mach. Learn. **106**(8), 1213–1241 (2017)
19. Sachpenderis, N., Karakasidis, A., Koloniari, G.: Structure and content based community detection in evolving social networks. In: Proceedings of the 11th International Conference on Management of Digital EcoSystems, pp. 1–8. ACM (2019)
20. Sachpenderis, N., Koloniari, G.: Determining interesting communities in evolving social networks. In: Proceedings of the 22nd Pan-Hellenic Conference on Informatics, pp. 249–254. ACM (2018)
21. Sadri, A.M., Hasan, S., Ukkusuri, S.V.: Joint inference of user community and interest patterns in social interaction networks. Soc. Netw. Anal. Min. **9**(1), 11 (2019)

22. Specia, L., Motta, E.: Integrating Folksonomies with the semantic web. In: Franconi, E., Kifer, M., May, W. (eds.) ESWC 2007. LNCS, vol. 4519, pp. 624–639. Springer, Heidelberg (2007). https://doi.org/10.1007/978-3-540-72667-8_44
23. Tennakoon, T., Nayak, R.: FCMiner: mining functional communities in social networks. Soc. Netw. Anal. Min. **9**(1), 20 (2019)
24. Toyoda, M., Kitsuregawa, M.: Extracting evolution of web communities from a series of web archives. In: Proceedings of the Fourteenth ACM Conference on Hypertext and Hypermedia, pp. 28–37. ACM (2003)
25. Wang, C.D., Lai, J.H., Philip, S.Y.: Neiwalk: community discovery in dynamic content-based networks. IEEE Trans. Knowl. Data Eng. **26**(7), 1734–1748 (2014)
26. Xie, J., Chen, M., Szymanski, B.K.: LabelRankT: incremental community detection in dynamic networks via label propagation. In: Workshop on Dynamic Networks Management and Mining, pp. 25–32 (2013)

A Weighted Feature-Based Image Quality Assessment Framework in Real-Time

Zahi Al Chami[1]([✉]), Chady Abou Jaoude[1], Bechara Al Bouna[1], and Richard Chbeir[2]

[1] TICKET Lab, Antonine University, Baabda, Lebanon
{zahi.chami,chady.aboujaoude,bechara.albouna}@ua.edu.lb
[2] University of Pau and Adour Countries, LIUPPA, Anglet, France
richard.chbeir@univ-pau.fr

Abstract. Nowadays, social media runs a significant portion of people's daily lives. Millions of people use social media applications to share photos. The massive volume of images shared on social media presents serious challenges and requires large computational infrastructure to ensure successful data processing. However, an image gets distorted somehow during the processing, transmission, sharing, or from a combination of many factors. So, there is a need to guarantee acceptable delivery content, especially for image processing applications. In this paper, we present a framework developed to process a large number of images in real-time while estimating the image quality. Our quality evaluation is measured based on four methods: Perceptual Coherence Measure, Semantic Coherence Measure, Content-Based Image Retrieval, and Structural Similarity Index. A weighted quality method is then calculated based on the four previous methods while providing a way to optimize the execution latency. Lastly, a set of experiments is conducted to evaluate our proposed approach.

Keywords: Image quality assessment · Real time data processing · Image functions adaptation

1 Introduction

With the recent advances in technology, data providers are continuously producing and streaming a significant amount of data as part of many scenarios, including news-feeds, pod-casts, and live interviews. More particularly and due to the growth of social media photo sharing applications, most of the data that are shared between users are images. As of June 2019, and according to [1], there are over than 300 million photos uploaded to Facebook every day, while more than 95 million photos are uploaded daily on Instagram.

This work is jointly funded from the National Council for Scientific Research in Lebanon (CNRS-L), the Antonine University, and the Agence universitaire de la Francophonie (AUF).

© Springer-Verlag GmbH Germany, part of Springer Nature 2020
A. Hameurlain et al. (Eds.) TLDKS XLV, LNCS 12390, pp. 85–108, 2020.
https://doi.org/10.1007/978-3-662-62308-4_4

These streams might be subject to alteration and/or modification; applying adaptation and/or protection functions to address the business needs. For example, blurring the face of an individual in an interview to hide his/her identity, removing sensitive content from a twitter stream, or highlighting only the objects of interest in a video due to some limitations in the hardware or in the connectivity. Hence, the outcome of content protection and/or adaption functions must be evaluated to ensure that the trade-off between the quality of the delivered content and the expected outcome is acceptable. For instance, it is important to evaluate the quality of the visual features in images prior to publishing. This can be done with structure content evaluation (e.g., ensuring that the objects of interest remained intact after applying content adaptation functions), and semantic content evaluation (e.g., ensuring that some useful information can still be extracted after applying content protection functions). Traditional approaches of evaluating and assessing data content do not scale well for data streaming scenarios due to the huge number of data that is being received and generated at tremendous rate.

These traditional processing techniques need first to store data before processing them, which will take a significant amount of time. However, and for fast-paced organizations, some quality-related decisions must be taken in real-time. To clarify the previous points, we provide the following scenario.

1.1 Motivating Scenario

Let us consider, for example, the case of a photo-sharing company (shown in Fig. 1) that provides its users with the ability to share and publish images online. Therefore, the company is demanded to process these images instantly as it receives an unbounded stream of images from its users.

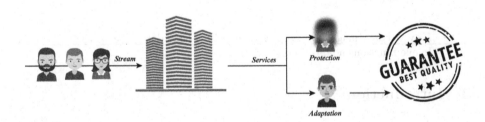

Fig. 1. Scenario

Moreover, the company offers its users some services such as:

- Protection to hide his/her identity using several techniques. For example: masking functions.
- Adaptation to satisfy some constraints imposed by the available resources, for example, image compression.

However, and after applying these services, some users complain about quality degradation. So, the company starts searching for a new solution to guarantee the delivered content quality as it is considered an important aspect. To sum up, the framework to be adopted by the company must be able to:

- Preserve the quality of the images: guarantee that the useful images' information and features such as color, shape, structure, etc. are kept intact and can still be extracted from the modified images in the output.
- Handle the unbounded stream of images: find a way to treat the significant number of images and ensure successful image processing instantly and in real-time, even though these services, along with the quality assessment process could take much time.

Several existing works focused on solving these challenges using many methods/techniques, which are cited in the next section along with their limitations before presenting our proposed approach.

2 Related Work

In this section, we present some existing works that are compared to our approach based on many criteria: 1) The remaining usefulness images' information when addressing business or users' needs, such as adaptation/or protection, 2) The methods used to assess the images' quality; this will show the number of features taken into consideration to estimate the quality, and 3) The time needed to process the images, especially the provided solutions that work in real-time.

Several techniques [8,15,22,29,30] modify/hide the features in the images to preserve privacy through content protection. Identity masking techniques [15,22,29,30], for instance, use black boxes (face hiding), large pixels (pixelation), blurring, swapping, and impainting, to obscure the whole or a portion of individual's face and/or body. Alternatively, content-based adaptation techniques [13,16,18] use frame dropping, semantic coding integration or compression to adapt multimedia content to the needs and interest of the viewers. Other techniques such as [11] and [24] build adaptation engines that are consistent with human visual perception and propose to improve the quality of the adapted video and to enhance the user's multimedia experience. These techniques tend to overlook the quality of multimedia content to achieve the adaption or protection.

Some existing studies involving the determination of users on a preferential adaptation of consumed multimedia contents can be seen in the literature. They can be categorized into two groups of utility: maximization-based and mathematical-based methods. In the former manner, the high-level goal is defined by a utility function of which optimal solution is the maximization of utility associating with concerned attributes in the user preferences. In the latter method, the high-level goal is instead considered as the decision objective of a multi-level hierarchical system structure of attributes and alternatives, based on their relationship. The problem resides when the utility maximization-based method exposes weaknesses on the demand of equivalent utility functions [19] for each attribute in the user preferences. Add to this, the knowledge required to solve the optimization problem.

In [25], the authors present a compression method to preserve the image quality while in [10,21], they used qualitative methods to assess the outcome quality of a distorted image in real-time. In the first one, they offered a machine learning-based approach to lossy image compression by producing files 2.5 times smaller than JPEG and JPEG 2000 while preserving its quality. In the second one, they provide a method to estimate the quality using a Perceptual Coherence Measure and assess the structure feature through Structural Similarity Index [31]. Despite the fact that these techniques give important results, they evaluate the distorted images by assessing only the color and structure features without considering the other features that might be altered and lead to quality degradation. The authors in [28] proposed a quality assessment method in live video streaming. Their method is divided into two main parts: 1) An offline deep unsupervised learning processes are employed at the server side, and 2) Inexpensive no-reference measurements at the client side.

They showed good results by comparing their metric to the FR benchmark using the Root Mean Square Error (RMSE). However, and despite processing the videos in real-time, they did not consider the execution latency and complexity, especially that they are working in real-time. The proposed approach in [14] specifies semantic constraints for video adaptation by defining a *utility function* to determine the utility of adaptation operators. This *utility function* is based on: (i) affected area, (ii) affected priority area, and (iii) the visual coherence of processed videos. In [9], the authors propose a framework that provides end-to-end quality control on real-time multimedia applications over heterogeneous networks. Their approach is based on a combined control of video assessment, Quality of Service (QoS), QoE-based mapping, and adaptation procedures. In [6], the authors present a framework of adaptive multimedia learning service where their engine allows the users to determine the best combination of adaptive features of video and audio content, under various constraints of network condition and user's context and preferences. While their approaches [6,9,14] provide useful insights to video adaptation, it fails to cope with content protection for two main reasons: 1) Adaptation operators differ from protection operators in a content processing perspective. For instance, dropping a scene can be interpreted as being omitted while scene blurring cannot, and 2) Using *gaps* to measure the visual coherence is not appropriate in content protection. For instance, blurring a face gives enough clues that could preserve semantic coherence, which is not the case for a drop operation.

We present, in Table 1, a summary of the previous cited approaches:

In our work, we aim to find a fair trade-off between the quality of the altered content and the expected outcome from the users in real-time as detailed in the next section.

A Weighted Feature-Based Image Quality Assessment Framework   89

Table 1. Showing the content and the limitations of each approach

Cited approaches	Content	Limitations
[8, 15, 22, 29, 30]	Modify/hide the features in the images to preserve privacy through content protection such as black box, large pixels, etc.	They overlook the quality of multimedia content to achieve the adaption or protection
[11, 13, 16, 18, 24]	They used techniques and adaptation engines that are consistent with human visual perception and propose to improve the quality of the adapted video	They rely on the PSNR metric to assess the quality, which is not enough to determine the extent of features degradation
[10, 21, 25]	They used qualitative methods such as SSIM, Perceptual Coherence Measure, etc. to assess the image quality	These approaches disregard the assessment of the remaining images' features, for example: color, shape, and texture.
[6, 9, 14]	They specify semantic constraints and provide utility function and Quality of Service to determine the utility	The techniques used are not appropriate in content protection as they are not preserving some useful information

3 Contributions

In this paper, we extend our previous work [10] by proposing a weighted average feature-based adaptive faces quality estimation approach for in-depth analysis of *perceptual coherence* and *features image assessment* in real-time. We assume that the faces, which are contained in the images, are affected by adaptations or protection functions to be evaluated on the fly.

Our contributions can be summarized as follow:

- We present a data-model representation for image content and features. We also provide in our model a set of data manipulation functions.
- We propose a weighted average feature-based adaptive faces quality estimation involving Perceptual Coherence and features assessment using Structural Similarity Index [31], Content-Based Image Retrieval [17], and Semantic Coherence to determine the image quality degradation. We adapted the general form of the referenced metrics according to our data model. Moreover, it is the first time that a combination of these metrics is taken into account while providing the possibility to choose and preserve specific features (depending on the users' needs and constraints) by assigning them a higher weight.
- We design a framework with an ability to efficiently evaluate a stream of images while providing a method to optimize the execution latency.

The remainder of this article is organized as follows. Section 4 presents some definitions and terminologies used in our work. The data quality assessment functions are described in Sect. 5, while the proposed framework is then detailed in Sect. 6. We evaluate our proposed approach in Sect. 7 through a set of experiments. Conclusions and future work are summarized in Sect. 8.

4   Definitions

In this section, we present the basic concepts needed to fully understand the proposed framework. We start by defining our data model and data manipulation functions that are used and needed later on in our framework.

4.1   Data Model

Definition 1 (image). *An image denoted by* **im** *is a basic data structure containing attributes that give information about its content. It is represented as follows:*

$$im \prec DESC, F, SO \succ$$

where,

- **DESC** *is a set of textual description, keywords or annotations provided by the user.*
- **F** *is the set of features that describes an image. It can be used to describe an entire image (global features) or a feature present at a location in the image space (local feature).*
- **SO** *is a set of salient objects that represent the objects of interests in an image as defined in the next section.*

Definition 2 (salient object). *A salient object designed by* **so** *represents an object of interest in the image, such as a person's face. It is defined as:*

$$so \prec w, h, coord, DESC, F \succ$$

where,

- **w** *and* **h** *are the width and height of the salient object so.*
- **coord** *represents the coordinates to determine the location of the salient object in the image.*
- **DESC** *is the set of annotations associated with the salient object.*
- **F** *is the set of features describing the visual content of a salient object (such as color distribution and intensity).*

Definition 3 (entity). *An entity expressed as* **e** *is a semantic object that exists by itself (e.g., person, vehicle). Each entity is associated with a set of salient objects. This association,* $e \rightarrow \{so_1, ..., so_n\}$*, done via manual or automatic annotation, highlights the salient objects* $\{so_1, ..., so_n\}$ *that are related to the entity e.*

Definition 4 (multimedia data stream). *A multimedia data stream denoted by* **mds** *is an infinite sequence of images formally defined as follows:*

$$mds = im_1, im_2, ..., im_k \text{ where } k \in \mathbb{N}^*$$

4.2   Data Manipulation Functions

In this section, we define the functions used to manipulate the multimedia data stream; either protect the salient objects or perform an adaptation on the multimedia data stream. In our assumptions, we focus mainly on identifying the salient objects that are subject to adaptation or protection. This goes beyond the rules defined under an authorization or adaptation scheme, which, for now, is beyond this paper's scope. We only consider that the functions used to protect or adapt the content are known and can be called implicitly on a subset of specified entities or images.

Definition 5 (image manipulation function). *An image manipulation function denoted by **imf** is a low-level function that alters, hides or deletes a set of features assigned to a salient object in an image im. $imf(so, im)$ takes a salient object so, the image in which so is contained and returns a modified salient object denoted by **so'**.*

As previously mentioned, we focus on two types of functions: a protection function and an adaptation function. As for the first type, it is used to suppress the content of an image by removing some of its features in order to protect an entity. For instance, as stated in [5], various techniques replace the salient objects' content with a manipulation function like a black box. Other techniques use face-swapping [15, 23] to choose a random or a default avatar to use in order to replace the content to be masked. The second type can be used to perform adaptation operations to meet resource constraints. The adaptation function modifies the image content in order to meet some hardware and software requirements. For example, in [7] an adaptation function is considered a robust video object cutout technique, which also can be called matting techniques, where a foreground object is pulled from a background image [12].

A combination of manipulation functions is applied on the set of images containing an entity. This combination is termed, in our approach, entity manipulation function and it is formally defined as follows;

Definition 6 (entity manipulation function). *An entity manipulation function denoted by emf is defined as:*

$$emf(e, mds) = (imf_1(so_1, im_1) \circ ... \circ imf_i(so_n, im_n))$$

*where, i and n $\in \mathbb{N}^*$. emf combines a set of image manipulation functions $(imf_1(so_1, im_1), ..., imf_i(so_n, im_n))$, which alters the salient objects associated with the entity e in the multimedia data stream mds, by modifying their features. As a result, emf(e, mds) returns a set of modified salient objects **SO'** associated with e.*

4.3   Image Quality Methods Background

Structural Similarity Index. The Structural Similarity Index (SSIM) [31] is a perceptual metric that quantifies image quality degradation. It is a full reference metric that requires as inputs two parameters: the modified salient objects SO′ with respect to a reference salient objects SO to quantify their visual similarity in image im. The general form of the SSIM index is defined as follows:

$$SSIM_{im}(SO, SO') = [l(SO, SO')]_\alpha \cdot [c(SO, SO')]_\beta \cdot [r(SO, SO')]_\gamma \qquad (1)$$

- α, β and γ are parameters to define the importance of each component.
- l(SO, SO′) index is related with luminance differences.
- c(SO, SO′) index is the contrast differences.
- r(SO, SO′) index is the structure variation.

These three indexes are computed as follows:

$$l(SO, SO') = (2\mu_{SO}\mu_{SO'} + C1)/(\mu_{SO}^2 + \mu_{SO'}^2 + C1)$$

$$c(SO, SO') = (2\sigma_{SO}\sigma_{SO'} + C2)/(\sigma_{SO}^2 + \sigma_{SO'}^2 + C2)$$

$$r(SO, SO') = (\sigma_{SOSO'} + C3)/(\sigma_{SO}\sigma_{SO'} + C3) \text{ where}$$

- μ_{SO} and $\mu_{SO'}$ are the average pixel values.
- C1, C2 and C3 are constants to avoid instabilities when $(\mu_{SO}^2 + \mu_{SO'}^2)$, $(\sigma_{SO}^2 + \sigma_{SO'}^2)$ or $\sigma_{SO}\sigma_{SO'}$ is equal to zero.
- σ_{SO} and $\sigma_{SO'}$ are the pixel value standard deviation.

We note that the score of the $SSIM_{im}(SO, SO')$ ranges from 0 (completely different) to 1 (identical images).

Content Based Image Retrieval. CBIR or Content-Based Image Retrieval [17] is the retrieval of images based on visual features such as color, texture, and shape. A CBIR architecture shown in Fig. 2 involves two steps:

- Feature extraction: The first step in the process is to extract image features to a distinguishable extent.
- Matching: The second step involves matching these features to yield a visually similar result.

CBIR systems extract features (color, texture, and shape) from images (original image) in the dataset based on the value of the image pixels. Each image stored in the dataset has its features compared to the features of the modified image. These features are smaller than the image size and stored in a database called feature database. Thus the feature database contains an abstraction (compact form) of the images in the image database; each image is represented by a compact representation of its contents (color, texture, shape) in the form of a fixed length real-valued multicomponent feature vectors or signature.

After forming the feature vectors, CBIR uses an image distance measure to compare the similarity of the modified image with those stored in the database in various dimensions and based on the visual content. In our work, we will use the Manhattan distance because, as stated in [26], it gives the best-retrieved result. For example, if u = $(x_1, x_2, ..., x_n)$ and v = $(y_1, y_2, ..., y_n)$ are two feature

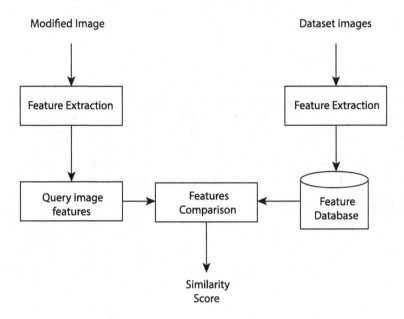

Fig. 2. CBIR architecture

vectors representing the modified image and a database image, the distance between these two vectors is given by:

$$\mathcal{D} = \frac{d_{MH}(u, v)}{n} \tag{2}$$

where:

$$d_{MH}(u, v) = \sum_{i=1}^{n} |x_i - y_i| \tag{3}$$

Equation (3) will return a score for each existing image in the database. We divided the value by n in order to normalize the result between 0 and 1. A distance of 0 indicates an exact match between two images, while the value 1 means that the two images are completely different with respect to the features that were considered. So the least the distance is, the more similar the images are. Then, the database image that has a smaller distance will be considered a match for the modified image. Then, the CBIR method is represented as follows:

$$CBIR_{im}(SO, SO') = 1 - \mathcal{D} \tag{4}$$

Where:

- SO is the original set of salient objects.
- SO' is the modified set of salient objects.

Finally, we subtracted the distance $\mathcal{D}$ from 1 in order to adjust the CBIR measure with the remaining quality methods.

5   Data Quality

5.1   Perceptual Coherence

Determining perceptual coherence consists of measuring the affected areas of images related to a specific entity. This is the essence of most techniques that use visual coherence [27]. However, and unlike the latter, our perceptual coherence measure computes the size of the distorted salient objects over the global image size. The premise of this assumption is that visually altering images covering relatively small affected areas would ultimately increase the quality of the images. We formally define the perceptual coherence measure as follows;

Definition 7 (Perceptual Coherence Measure). *A Perceptual Coherence Measure denoted by $PCM_{im}(SO')$ quantifies the affected area in im based on the relative size of the salient objects, where SO' are the set of modified salient objects that represent the entities in an image im. The perceptual coherence measure of the set of salient objects SO', $PCM_{im}(SO')$ is computed as follows;*

$$PCM_{im}(SO') = 1 - \frac{(\sum\limits_{so' \in SO'} \int\limits_{so'.coord_x}^{so'.coord_x + so'.h} \int\limits_{so'.coord_y}^{so'.coord_y + so'.w} dx\, dy) - \underset{i \in 1...n-1, j \in i+1...n}{IntersectArea} (so'_i, so'_j)}{sizeof(im)}$$

$$where$$

$$IntersectArea(so'_i, so'_j) = (min(x_{i_{TopRight}}, x_{j_{TopRight}}) - max(x_{i_{BottomLeft}}, x_{j_{BottomLeft}}))*$$

$$(min(y_{i_{TopRight}}, y_{j_{TopRight}}) - max(y_{i_{BottomLeft}}, y_{j_{BottomLeft}})) \tag{5}$$

$$with : \begin{cases} x_{i_{TopRight}} & = so'_i.coord_x + so'_i.w \\ x_{j_{TopRight}} & = so'_j.coord_x + so'_j.w \\ x_{i_{BottomLeft}} & = so'_i.coord_x + so'_i.h \\ x_{j_{BottomLeft}} & = so'_j.coord_x + so'_j.h \\ y_{i_{TopRight}} & = so'_i.coord_y + so'_i.w \\ y_{j_{TopRight}} & = so'_j.coord_y + so'_j.w \\ y_{i_{BottomLeft}} & = so'_i.coord_y + so'_i.h \\ y_{j_{BottomLeft}} & = so'_j.coord_y + so'_j.h \end{cases}$$

In this equation, we are taking into consideration all the modified salient objects representing each entity in an image. To do so, we measure the area of each salient object, which is represented by a rectangle that has coordinates. While calculating the area, we need to ensure that:

– At least one salient object, which represents a specific entity, should exist.
– If an intersection area exists between two or more salient objects, this area will be only calculated once.

The numerator will return the total area size of the salient objects in the image. The result is divided by the image area size, which is indicated by sizeof(im), to obtain the ratio representing the modified area relative to the entire space. Lastly, we subtract the ratio from 1 to find the coherence value. We note that there are no modified salient objects if the equation is equal to 1. The size of the area that is affected by a manipulation function plays an important role in the perceptual coherence.

5.2 Semantic Coherence Measure

Determining the Semantic Coherence Measure consists of assessing the facial expression features related to a specific entity. Facial expressions give important clues about emotions. In our work, we considered seven expressions as the main emotional expressions that are common among human beings. Let EXP be a vector that contains the set of expressions. It is written as follows:

$$EXP = [anger, disgust, fear, happiness, sadness, surprise, neutral]$$

We denoted by p the probability of each emotion, and the sum of these emotions' probabilities is equal to 1. To clarify the emotion recognition process, we are considering Fig. 3 shown below.

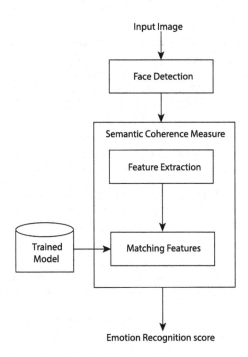

Fig. 3. Face emotion recognition

In most existing facial expressions datasets, each facial image is only associated with one single emotion, but a face may have multiple emotions.

For example, someone can be happily surprised or angrily disgusted. However, and in our work, we are choosing the emotion that has the highest probability. The emotions are recognized through the use of a machine learning pre-trained model and it involves two steps:

- Detecting frontal face in an image.
- Comparing the faces to a pre-trained CNN model architecture which takes the bounded face as input and predicts probabilities of the seven emotions, which are listed before, in the output layer.

We first apply an emotion recognition formula before calculating the Semantic Coherence Measure, denoted by $ER_{im}(SO)$, with SO the set of salient objects representing the entities in image im. The formula is computed as follows:

$$ER_{im}(SO) = \frac{\sum_{so \in SO} \max_{i \in 0...6} p(EXP(i))}{count(SO)} \qquad (6)$$

In formula (6), we are calculating the emotion probability of the salient objects in im and dividing the result by the number of salient objects in order to maintain the value between 0 and 1.

Definition 8 (Semantic Coherence Measure)
A Semantic Coherence Measure denoted by $SCM_{im}(SO, SO')$ quantifies the quality of an image im by assessing the facial expression features between the modified salient objects SO' and the original one SO. The Semantic Coherence Measure, $SCM_{im}(SO, SO')$ is computed as follows in order to estimate the semantic loss;

$$SCM_{im}(SO, SO') = 1 - \frac{|ER_{im}(SO') - ER_{im}(SO)|}{ER_{im}(SO)} \qquad (7)$$

This formula will return between 0 and 1. Higher scores mean face emotion preservation.

5.3 Image Score

Lastly, and in order to find the final image quality score, we are combining the four previous methods. *Image Score*, denoted by IS, is computed as follows:

$$IS = \frac{w_1 * [PCM_{im}(SO')] + w_2 * [SSIM_{im}(SO, SO')] + w_3 * [SCM_{im}(SO, SO')] + w_4 * [CBIR_{im}(SO, SO')]}{\sum_{i=1}^{4} w_i}$$
$$(8)$$

where w_1, w_2, w_3 and w_4 are weights between 0 and 1 with their sum equal to 1. These weights are chosen by the administrator to indicate the importance of each method. We note that the score of IS ranges from 0 to 1. Higher scores indicate quality preservation and similarity between the original image and the modified one.

6   Proposed Framework

An overview of our framework is shown in Fig. 4. It consists of two main modules:

– Stream Processing.
– Back-end.

In the following, we present in details the framework's modules.

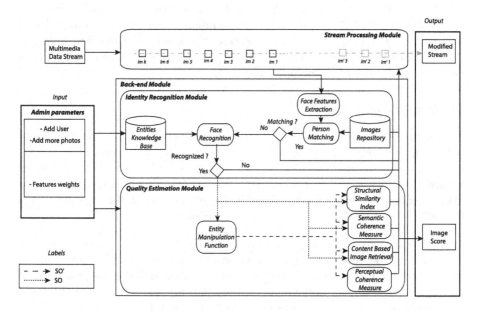

Fig. 4. Framework

6.1   Stream Processing Module

In this module, administrators query continuous data streams and detect conditions within a small time period from the time of receiving the data. In our work, we took Twitter as a source of multimedia data stream while processing only images.

As shown in Fig. 4, the images contained in the stream are marked from one to k where k is equal to an infinite number in order to indicate that we are treating images without any bound. In the end, im′ is the resulted images that are returned from the back-end module.

6.2   Back-End Module

It consists of two main submodules: a) Identity Recognition and b) Quality Estimation.

The first one is responsible of:

- Detecting and recognizing the entities.
- Checking if an incoming image is already processed in order to minimize the execution time.

These previous works are done through the use of five components: 1) *Face Features Extraction*, 2) *Images Repository*, 3) *Person Matching*, 4) *Face Recognition* and 5) *Entities Knowledge Base*.

The second submodule has the duty to assess the image quality and find its score with the aid of the remaining components: 6) *Entity Manipulation Function*, 7) *Perceptual Coherence Measure*, 8) *Structural Similarity Index*, 9) *Content Based Image Retrieval* and 10) *Semantic Coherence Measure*. We will detail each component in the upcoming sections.

Identity Recognition Module

Face Features Extraction: Face Features Extraction is a process of dimensionality reduction by which an image is reduced to more manageable groups for processing. It is useful in order to reduce the number of resources and the time needed for processing without losing important information. We focused in our work on extracting features related to the faces. These features will form a feature vector that will be sent to the *Image Matching* component.

Images Repository: This is a container where the processed images are stored. To minimize the time needed to search for a possible match, we are clustering and grouping the images based on the person's identity. So, the number of groups are created from the number of users that are trained and chosen by the administrator. But, due to the huge amount of images that might be saved, we are addressing this issue by applying the following points:

- The Least Frequently Used (LFU) algorithm [20] that uses a counter to keep track of how often an image is accessed while retaining only the feature vectors of each image to reduce the image size and to facilitate the search process. With LFU algorithm, the image with the lowest count is removed first.
- The administrator has the ability to specify the number of images referred to each person within each cluster.

Moreover, and after performing the grouping of the images, each group will be assigned an average feature vector that describes the persons' identity.

Person Matching: This component is responsible for comparing the features of the persons within the images that are coming from the stream with those stored in the *Images repository*. If a match is found for all the persons contained in *im*, the image will return to the stream processing module while assigning this image the quality score of the matched persons that exist in the *images Repository*. Otherwise, it will continue its process by recognizing the remaining unknown persons, which are not found in the *Images Repository*, using the next component.

Face Recognition: If no match is found in the previous component and in order to identify the entities, we used the extracted features (from the *Face Features Extraction* component) that will be compared with those stored in the *Entities Knowledge Base*, which will be detailed in the next section, using the *Face Recognition* component. If a match is found, the image is forwarded to the *Entity Manipulation Function*. Otherwise, the image will be directly returned to the *Stream Processing* module.

Entities Knowledge Base: This component represents the database where the trained entities reside. An administrator has an option to add more entities to the database in order to create his own schema and to train more photos to the existing entities as well. This will offer the opportunity to improve the recognition accuracy.

Quality Estimation Module

Entity Manipulation Function: This component modifies the salient objects of the entities, which are considered the faces in our work. This function will take the entities as inputs and apply a set of image manipulation functions on the salient objects representing these entities. As a result, it will return a set of modified salient objects.

As we mentioned before, the image manipulation function can be either a protection or adaptation function. We used three main protection functions: pixelate, gaussian blur and median blur. Concerning the adaptation function, we used two compression techniques: lossy and lossless. In fact, the functions differ by means of the features that they preserve. For instance, a median blur is a function that returns a modified image in such a way that some of the visual, semantic and multimedia features are hidden while metadata and audio features remain intact. Giving that each function preserves certain features, we are applying here a list of functions in order to find the most appropriate one that will guarantee an acceptable image quality through the use of the quality assessment functions.

Perceptual Coherence Measure: After modifying the faces, the first metric used to assess the image is the visual coherence measure through the use of the PCM. In this component, we measure the total area of the modified faces. As a result, the function will return a value between 0 and 1 that will be aggregated with the next quality methods.

Structural Similarity Index: The second metric used to estimate the image quality in our framework is the *Structural Similarity Index*. It will return a score for each manipulation function. Then, this component will help us to select the manipulation function that has the highest score for the modified image in terms of structure, luminance and contrast.

Semantic Coherence Measure: This component has the responsibility of determining the image quality by assessing the face emotional state based on facial expressions. It will also return a score between 0 and 1 for each manipulation function. Higher scores indicate facial expressions preservation that will lead to conserving the face emotion.

Content Based Image Retrieval: The last metric adopted in our framework to assess the image quality is the *CBIR*. It will evaluate the quality in terms of color, shape and texture. As a result, a score will be returned between 0 and 1, which will indicate the extent of the features degradation.

Image Score: The final component is the *Image Score* that is responsible for:

- Aggregating the scores, which are returned from the previous quality methods, based on their importance. An administrator has the privileges to select his preferred features by assigning them parameters, which are known as features weights.
- Displaying the final image score.

Simultaneously, the modified image (im′) will return to the *stream processing module* in order to be then published.

7   Experiments

In order to test the efficiency of our approach, a program is developed by java language using eclipse on a desktop computer with a 2.66 GHz core 2 duo and 4 GB RAM running Linux Ubuntu 14.04 64 bit. After testing the program on one desktop, the framework described above is implemented on a distributed environment called Apache Storm [2]. In order to successfully run the storm cluster, we must implement all of its components. For that purpose, we used 16 physical machines, along with the needed libraries, as shown in the Table below:

Table 2. Showing the Apache Storm Configuration and the needed libraries

Apache Storm Configuration	
Machine	Service
Client node [3]	It tests the framework locally before deploying it to the cluster
Nimbus node [4]	It deploys the framework, schedules the tasks to the workers' nodes, and monitors the progress of the tuples in the topology
Three Zookeeper nodes [4]	They handle the communication between the nimbus and the supervisors. Also, they keep the states of the nimbus and the supervisors
Eleven Supervisor nodes [4]	They monitor the workers by viewing the heartbeat of each one of them to determine its state. Each worker runs one or more worker processes, and each worker process runs a JVM that contains one or more executors
Implemented libraries	
Libraries' name	Service
Apache Storm 0.9.3 and zookeeper 3.4.6	They must be implemented on all nodes to successfully run the storm cluster.
OpenCV 3.4.3 API and python dlib library	We use the manipulation functions from OpenCV and performing face detection using dlib.
Pre-trained ResNet model	it is used to recognize the faces and their expressions

Note that all these computers are connected to each other using a switch and giving each computer a static IP address.

We conducted two sets of experiments. In the first one, we assessed the efficiency of our algorithm. More precisely, we evaluated the image data quality that may be reduced by applying a manipulation function. In this scenario, we limit the size of the processed images from Twitter Stream to 2100 as our goal is to determine the image quality using the PCM, SSIM, CBIR and SCM methods that are defined in Sect. 4. To do so, we started our scenario by varying the number of faces contained in the images from 1 to 3 and applying several manipulation functions to find the suitable one that will return the best score in terms of quality. Lastly, the prototype is tested only on a local cluster without being uploaded to the distributed system. In the second one, we evaluated the Apache Storm performance in terms of: 1) Execution latency: The average time an image spends during its execution, and 2) Number of nodes: Number of supervisors that are used to process the images.

To do so, the following scenario was executed:

1. We randomly selected 100 individuals and trained 50 photos for each person.
2. We have processed 50 000 images from Twitter Stream.
3. We distributed the libraries on all nodes.

4. Finally, we uploaded the framework to the cluster, by starting on two nodes as the number will be incremented by 2 in order to evaluate the performance of Apache Storm.

7.1 Test 1: Evaluation of Image Quality Affected by a Manipulation Function

The objective of this test is to evaluate the image quality that may be affected when applying a manipulation function. We use in this study three manipulation functions, and they are mainly considered as protection functions: Pixelation (a.k.a mosaicking), Gaussian blurring, and Median blurring. We specify for each manipulation function fixed parameters while allowing the users to choose a weight for each quality method based on their preferred image features. The manipulation functions parameters, as well as the quality methods weights, are shown in Table 3.

Table 3. Parameters list

Manipulation functions parameters list			
Manipulation functions	Gaussian blur	Median blur	Pixelate	–
Kernel size	31 × 31	31 × 31	–	–
standard deviation	5	5	–	–
Pixel Size	–	–	10	–
Quality methods weights			
Quality methods	PCM	SSIM	CBIR	SCM
Weights	0	0.6	0.2	0.2

We chose the manipulation functions parameters according to the literature as they are considered an average intensity values for each manipulation function. In order to test the efficiency of this method, we process 2100 images from Twitter Stream as they are divided into three parts based on the number of persons (from one to three persons) contained in each image. We then apply the manipulation functions on each image, and as a result, we obtained a Table and a graph shown in Table 4 and Fig. 5. These results represent the SSIM, SCM, CBIR, and PCM average values for each manipulation over 2100 images.

According to the above graph (Fig. 5), the manipulation function that has the highest image score is the Gaussian blur. Moreover, and based on Table 2, this function will satisfy the needs of the users because it preserves the structure, contrast, emotion, and color features that are chosen by the users and assessed by the SSIM, SCM, and CBIR methods.

Table 4. Quality scores when applying a manipulation function, where * represents the number of persons

Quality scores before and after adding weights for each manipulation function						
Manipulation functions	Gaussian blur		Median blur		Pixelate	
Applying weights	Before	After	Before	After	Before	After
SSIM 1*	0.9894	0.5936	0.9875	0.5925	0.9832	0.5899
SSIM 2*	0.9886	0.5931	0.9865	0.5919	0.9815	0.5889
SSIM 3*	0.9904	0.5942	0.9890	0.5934	0.9849	0.5909
CBIR 1*	0.9878	0.1975	0.9856	0.1971	0.9908	0.1981
CBIR 2*	0.9885	0.1977	0.9863	0.1972	0.9907	0.1981
CBIR 3*	0.9894	0.1978	0.9883	0.1976	0.9920	0.1984
SCM 1*	0.2480	0.0496	0.1552	0.0310	0.1967	0.0393
SCM 2*	0.2388	0.0477	0.1614	0.0322	0.1905	0.0381
SCM 3*	0.2140	0.0428	0.1625	0.0325	0.1816	0.0363
PCM 1*	0.9783	0	0.9783	0	0.9783	0
PCM 2*	0.9611	0	0.9611	0	0.9611	0
PCM 3*	0.9541	0	0.9541	0	0.9541	0

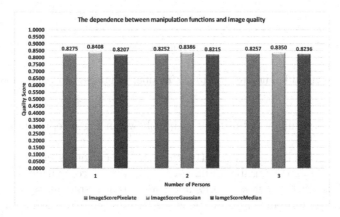

Fig. 5. Showing the dependence of a manipulation function regarding the image quality score.

7.2 Test 2: Evaluation of Apache Storm Performance in Real-Time

In this test, we treat 50 000 images. But before starting our experiments, we randomly selected and trained around 5000 images representing 100 individuals. Two sets of experiments are carried out in order to evaluate the performance of our framework, which is measured from the Apache Storm distributed system by assessing the execution latency at each component. In the first one, we fixed the number of nodes to 4 and used various number of images ranging from 5000

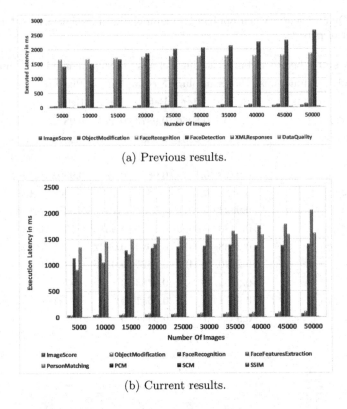

(a) Previous results.

(b) Current results.

Fig. 6. Executed latency related to the number of images at each component.

to 50 000. As a result, we obtain the graph shown in Fig. 6 representing the first study.

We noticed in Fig. 6(a) and 6(b) that the execution latency will increase while incrementing the number of images. Moreover, by comparing the execution latency between the previous results and the current results at each component, we can see a performance improvement concerning the execution latency.

In our second study, we fixed the number of images to 50 000 and took a distinct number of nodes, starting from 2 to 7. Therefore, the results are shown in Fig. 7 and 8.

According to the above graph (Fig. 7), we can clearly see an improvement in terms of the execution latency at each component due to the fact that some of the persons contained in the images are already processed and stored in our images repository, which will allow us to bypass the remaining components.

We notice that while increasing the number of nodes, an improvement in Apache Storm performance can be noted as the time needed to execute these images will decrease for the reason that the number of workers and executors is incremented, which may lead to handling more tasks at a time. In addition, we can note a decrease (or a little bit higher) in the execution time for the whole framework despite the fact that the number of the components has increased.

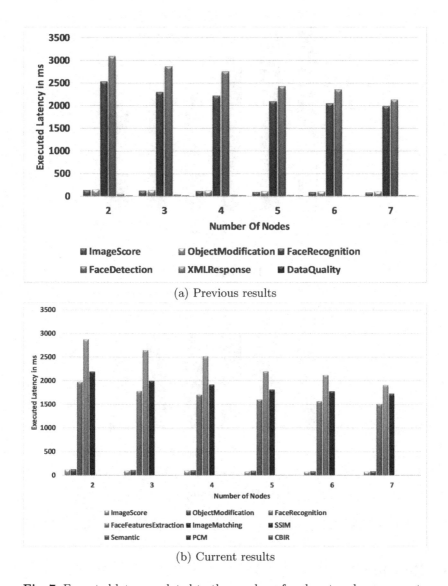

(a) Previous results

(b) Current results

Fig. 7. Executed latency related to the number of nodes at each component.

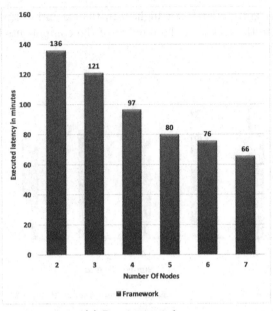

(a) Previous results.

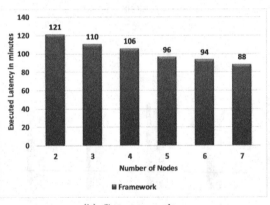

(b) Current results.

Fig. 8. Executed latency related to the whole Framework.

8 Conclusion

In this paper, we presented a framework with the intention of achieving an image quality estimation that may be distorted during the processing and transmission phase while treating these images in real-time. Our quality estimation is based on four methods: Perceptual Coherence Measure, Structural Similarity Index, Content-Based Image Retrieval and Semantic Coherence Measure. We then aggregated these methods in a weighted average quality estimation where the user has the ability to choose the weights based on his preferred features. We also provided a method to optimize the execution latency. A set of experiments has been tested in order to evaluate our approach. We noticed an improvement in the execution latency between our previous and current results.

Our current framework can only be applied to the Full-reference images. So, in future work, we intend to provide a neural network model that will be able to estimate the quality of the No-reference images by assessing several features.

References

1. Social media statistics in 2020. https://dustinstout.com/social-media-statistics/#instagram-stats. Accessed 24 Jan 2020
2. Apache storm - concepts (2015). http://storm.apache.org/releases/current/Concepts.html
3. Setting up a development environment (2015). http://storm.apache.org/releases/1.0.6/Setting-up-development-environment.html
4. Apache storm cluster architecture (2018). http://storm.apache.org/releases/1.0.6/Setting-up-development-environment.html
5. Agrawal, P., Narayanan, P.: Person de-identification in videos. IEEE Trans. Circuits Syst. Video Technol. **21**(3), 299–310 (2011)
6. Atchara Rueangprathum, S.L., Witosurapot, S.: User-driven multimedia adaptation framework for context-aware learning content service. J. Adv. Inf. Technol. **7**, 182–185 (2016)
7. Bai, X., Wang, J., Simons, D., Sapiro, G.: Video SnapCut: robust video object cutout using localized classifiers. In: ACM SIGGRAPH 2009 papers, SIGGRAPH 2009, pp. 70:1–70:11. ACM, New York (2009)
8. Al Bouna, B., Chbeir, R., Gabillon, A.: The image protector - a flexible security rule specification toolkit. In: SECRYPT, pp. 345–350 (2011)
9. Cerqueira, E., Fernando Boavida, A.M.: Quality of experience management framework for real-time multimedia applications (2009)
10. Chami, Z., AL Bouna, B., Jaoude, C., Chbeir, R.: A real-time multimedia data quality assessment framework, pp. 270–276 (2019). https://doi.org/10.1145/3297662.3365803
11. Kim, C.S., Sohn, H., De Neve, W., Ro, Y.M.: An objective perceptual quality-based ADTE for adapting mobile SVC video content. IEICE Trans. Inf. Syst. **92**, 93–96 (2009). Please check and confirm the edit made in author names in Ref. [11]
12. Chuang, Y.Y., Agarwala, A., Curless, B., Salesin, D.H., Szeliski, R.: Video matting of complex scenes. In: Proceedings of the 29th Annual Conference on Computer Graphics and Interactive Techniques, SIGGRAPH 2002, pp. 243–248. ACM, New York (2002)
13. De Bruyne, S., De Schrijver, D., De Neve, W., Van Deursen, D., Van de Walle, R.: Enhanced shot-based video adaptation using MPEG-21 generic bitstream syntax schema. In: IEEE Symposium on Computational Intelligence in Image and Signal Processing, CIISP 2007, pp. 380–385, April 2007. https://doi.org/10.1109/CIISP.2007.369199
14. El-Khoury, V., Bennani, N., Coquil, D.: Utility function for semantic video content adaptation. In: Proceedings of the 12th International Conference on Information Integration and Web-based Applications & Services, iiWAS 2010, pp. 921–924. ACM, New York (2010)

15. Fan, J., Luo, H., Hacid, M.S., Bertino, E.: A novel approach for privacy-preserving video sharing. In: Proceedings of the 14th ACM International Conference on Information and Knowledge Management, CIKM 2005, pp. 609–616. ACM, New York (2005)

16. Gang, Z., Chia, L.T., Zongkai, Y.: MPEG-21 digital item adaptation by applying perceived motion energy to H.264 video. In: 2004 International Conference on Image Processing, ICIP 2004, vol. 4, pp. 2777–2780, October 2004. https://doi.org/10.1109/ICIP.2004.1421680

17. Gudivada, V.N., Raghavan, V.V.: Content based image retrieval systems. Computer **28**(9), 18–22 (1995)

18. Herranz, L.: Integrating semantic analysis and scalable video coding for efficient content-based adaptation. Multimed. Syst. **13**, 103–118 (2007). https://doi.org/10.1007/s00530-007-0090-0

19. Kephart, J.O., Das, R.: Achieving self-management via utility functions. IEEE Internet Comput. **11**, 40–48 (2007)

20. Ketan, P., Anirban, S., Matani, M.D.: An O(1) algorithm for implementing the LFU cache eviction scheme (2010)

21. Li, Q., Lin, W., Fang, Y.: No-reference quality assessment for multiply-distorted images in gradient domain. IEEE Signal Process. Lett. **23**(4), 541–545 (2016). https://doi.org/10.1109/LSP.2016.2537321

22. Newton, E.M., Sweeney, L., Malin, B.: Preserving privacy by de-identifying face images. IIEEE Trans. Knowl. Data Eng. **17**(2), 232–243 (2005)

23. Nguyen, S.M., Ogino, M., Asada, M.: Real-time face swapping as a tool for understanding infant self-recognition. CoRR abs/1112.2095 (2011)

24. Prangl, M., Szkaliczki, T., Hellwagner, H.: A framework for utility-based multimedia adaptation. IEEE Trans. Circuits Syst. Video Technol. **17**(6), 719–728 (2007). https://doi.org/10.1109/TCSVT.2007.896650

25. Rippel, O., Bourdev, L.: Real-time adaptive image compression. In: Proceedings of the 34th International Conference on Machine Learning, ICML 2017, vol. 70, pp. 2922–2930. JMLR.org (2017). http://dl.acm.org/citation.cfm?id=3305890.3305983

26. Sural, S., et al.: Performance comparison of distance metrics in content-based image retrieval applications. In: Proceedings of the International Conference on Information Technology, Bhubaneswar, India, pp. 159–164, January 2003

27. Truong, B., Venkatesh, S., Dorai, C.: Scene extraction in motion pictures. IEEE Trans. Circuits Syst. Video Technol. **13**(1), 5–15 (2003)

28. Vega, M.T., Mocanu, D.C., Famaey, J., Stavrou, S., Liotta, A.: Deep learning for quality assessment in live video streaming. IEEE Signal Process. Lett. **24**(6), 736–740 (2017). https://doi.org/10.1109/LSP.2017.2691160

29. Vijay Venkatesh, M., Cheung, S.c.S., Zhao, J.: Efficient object-based video inpainting. Pattern Recognit. Lett. **30**(2), 168–179 (2009)

30. Wickramasuriya, J., Datt, M., Mehrotra, S., Venkatasubramanian, N.: Privacy protecting data collection in media spaces. In: Proceedings of the 12th Annual ACM International Conference on Multimedia, MULTIMEDIA 2004, pp. 48–55. ACM, New York (2004)

31. Wang, Z., Bovik, A.C., Sheikh, H.R., Simoncelli, E.P.: Image quality assessment: from error visibility to structural similarity. IEEE Trans. Image Process. **13**(4), 600–612 (2004). https://doi.org/10.1109/TIP.2003.819861

Sharing Knowledge in Digital Ecosystems Using Semantic Multimedia Big Data

Antonio M. Rinaldi[1,2(✉)] and Cristiano Russo[1]

[1] Department of Electrical Engineering and Information Technologies,
University of Napoli Federico II, Via Claudio, 21, 80125 Napoli, Italy
{antoniomaria.rinaldi,cristiano.russo}@unina.it
[2] IKNOS-LAB Intelligent and Knowledge Systems - LUPT,
University of Napoli Federico II, Via Toledo, 402, 80134 Napoli, Italy

Abstract. The use of formal representations has a basic importance in the era of big data. This need is more evident in the context of multimedia big data due to the intrinsic complexity of this type of data. Furthermore, the relationships between objects should be clearly expressed and formalized to give the right meaning to the correlation of data. For this reason the design of formal models to represent and manage information is a necessary task to implement intelligent information systems. Approaches based on the semantic web need to improve the data models that are the basis for implementing big data applications. Using these models, data and information visualization becomes an intrinsic and strategic task for the analysis and exploration of multimedia Big Data. In this article we propose the use of a semantic approach to formalize the structure of a multimedia Big Data model. Moreover, the identification of multimodal features to represent concepts and linguistic-semantic properties to relate them is an effective way to bridge the gap between target semantic classes and low-level multimedia descriptors. The proposed model has been implemented in a NoSQL graph database populated by different knowledge sources. We explore a visualization strategy of this large knowledge base and we present and discuss a case study for sharing information represented by our model according to a peer-to-peer(P2P) architecture. In this digital ecosystem, agents (e.g. machines, intelligent systems, robots,...) act like interconnected peers exchanging and delivering knowledge with each other.

Keywords: Semantic BigData · Multimedia ontologies · Semantics · P2P

1 Introduction

The huge amount of data produced every day by humans and machines in different formats and in several application areas has a great impact on methodologies and technologies able to capture, store and analyze data. The Big Data paradigm tries to give a comprehensive approach to deal with the management

© Springer-Verlag GmbH Germany, part of Springer Nature 2020
A. Hameurlain et al. (Eds.) TLDKS XLV, LNCS 12390, pp. 109–131, 2020.
https://doi.org/10.1007/978-3-662-62308-4_5

of extreme volumes of heterogeneous data and other dimensions as the high velocity of changing data [16]. The Internet has a crucial role in this context both in the creation and in the access to a large amount of online information sources, underlining the problem of managing data. However, web content is generally for human use and not for machine processing. The increasing volume, variety and complexity of the data collected presents interesting challenges in different research fields. In this scenario, both humans and machines must not only be able to access information, but also to interpret, use and share it. In our opinion, the use of semantics could be a useful way to achieve these goals by using approaches from different research fields such as semantic data integration or semantic data extraction. [13,17,57]. Our approach is strictly related to the vision of Semantic Web [8] voted to give formal tools to solve or at least smooth ambiguities, inconsistencies and heterogeneities at different data levels. A common standard for data format and structure is a useful tool in order to integrate heterogeneous models in a unified conceptualization of a specific knowledge or application domain and reuse of existing knowledge models [12], allowing intelligent systems to share knowledge in a real environment. Moreover, the W3C has proposed during years a number of standards to implement this vision as RDF, OWL, SPARQL and promoted the use of ontologies to give a formal, common and shared view of data. The union of the Big Data paradigm and the Semantic Web vision is a new and interesting research area called Semantic Big Data [32,35]. The combination of these approaches will define more efficient techniques for storing, organizing and analyzing the huge amount of available information, facilitating the work of data scientists reducing, for example, the information overload and redundancy of data. In addition, other models based on ontologies may be merged in new ones, linking information from different sources and making it possible to realize more efficient tools and smart applications [41,45]. While ontologies and big data paradigm are silver bullets for formal representation and efficient management of a large amount of data, we define an integrated, general purpose model to enable a more expressive way to represent data and transform it in information and knowledge. We use a property-based labeled graph to represent our model and implement it in a graph database. We decided to use a graph structure because we consider it as a natural representation of complex ontologies [15]. The graph database is populated using general multimedia knowledge bases and linked open data and we implement a system to share knowledge represented by our graph database in a P2P digital ecosystem.

The motivations behind this work are manifold. The need for semantic models able to elevate raw data and information to knowledge is one of the keys for the development of next future intelligent systems. This need is still more evident if we consider the rapid evolution of our society, where humans, robots and other machines are required to collaborate and share their knowledge with each other by means of friendly interactions. In this context, the use of a formal model, based on NoSQL technologies is a first step to take into account complex systems and to provide an adequate scalability of large-scale multimedia big data due to the intrinsic complexity of unstructured data. Furthermore, the inclusion of linguistic-semantic interconnections between model entities enables the addition of "interpretability" feature to the model, which is essential for a shared understanding of knowledge to be shared.

The reminder of the paper is organized as follows: in Sect. 2 is presented the state of the art of our field of interest; Sect. 3 is devoted to give a detailed description of our model, also providing insights about the implemented model in a NoSQL database. In Sect. 4 we show a case study for sharing the knowledge through a peer-to-peer architecture; eventually, in Sect. 5 conclusion and future works are discussed.

2   Related Works

In this Section we describe and discuss the directions of Big Data research during years, focusing our attention on the evolution process which involves semantics and ontologies for Big Data development. The use of the term "Big Data" in articles and reports is today overwhelming. The pervasive nature of digital technologies and the broad range of data-reliant applications have also made this expression widespread across other disciplines, including sociology, medicine, biology, economics, management and information science. However, this raising phenomenon has not been accomplished by a rational development of a shared vocabulary. In fact, the word has been used with several and inconsistent meanings and lacks of formal definitions. One of the first attempt for a stately definition is given in [23]. According to the authors, Big Data is an information asset characterized by a High *Volume*, *Velocity* and *Variety* to require specific technology and analytical methods for its transformation into *Value*. This model can be even extended to 5Vs if the concept of *Veracity* is incorporated into the big data definition[6]. A more articulated classification of big data literature regarding models, data types and applications can be found in [34]. While much attention has been devoted to the *Volume* and *Velocity* dimensions of Big Data, a systematic support for managing the *Variety* of data, is only emerging in recent years. In [35], this process is called "Semantification" of Big Data. In this paper the authors present an approach for managing hybrid Big Data using RDF data model, i.e. non-semantic Big Data is semantically enriched by using RDF vocabularies. In terms of data variety, this approach allows to ingest both semantic and non-semantic data. However, it is limited to convert non-semantic data to RDF data, where possible. In [27], a survey on Big Data concepts and challenges is given. It also discusses the problem of merging Big Data architecture in an already existing information system, tackling the emerging importance of semantics (reasoning, coreference resolution, entity linking, information extraction, consolidation, paraphrase resolution, ontology alignment) in the Big Data context. The use of semantics for big data processing has had an important increasing in many applications fields. In [1] the authors present a distributed architecture and technology for scaling up text analysis running a complete chain of linguistic processors on several virtual machines. This approach is limited to textual analysis and does not consider multimedia data. An approach based on semantics and ontologies is also a key aspect in "social big data" [6] and in multimedia social networks [19]. In [38] a semantic big data platform to integrate heterogeneous wearable data in healthcare has been presented. Semantic big data is also used for realizing a national-scale infrastructure for vulnerability analysis in critical infrastructure systems (CIs) such as energy, water, transportation, and communication [33]. In [55] Knowle, an online news management system based on semantic link network model is described. Another system exploiting semantics is *Karma* [32]; it is a system applied in the cultural heritage domain to integrate data across museums. Above-cited approaches have the advantage of using semantics, but their models have been designed for very limited and specific knowledge domains. A semantic ETL framework is proposed in [4]. It generates a semantic model to integrate heterogeneous datasets, and then generates semantic linked data complied with the data model. The generated semantic data is made available on the Web as linked data (RDF triples) that can be queried and used in analytic tasks and as a resource for innovative data-driven applications. In [7] an approach handling semantic heterogeneity and URI-based entity identification over multiple data sources is proposed. The work describes a semantic entity resolution method based on inference mechanism using rules to manage the misunderstanding of data, for real world entities; a Data Quality enhancement using MapReduce-based query rewriting approach; a parallel combination of MapReduce jobs and query rewriting inferences to handle transitive and cyclic rules for a richer rule expression language. A scalable data integration platform is presented in [10]. It uses Big Data technologies based on ontological models to give semantic-based analysis services for various purposes. Semantic technology, or the use of ontologies, is seen as a core approach to solve the big data *Variety* challenge and align the data generated from heterogeneous data sources as discussed in [47] with a special focus on the financial domain. An approach based on an unsupervised and adaptive ontology-learning process is described in [30]. The growing amount of information from the Web is processed and extracted to get only the most valuable pieces of information. The resulting ontology is then used to enhance the performance of a focussed crawler. The combination of Big Data and Semantic Web technologies allows to classify information according to a domain knowledge. MOUNT [48] is a multi-level annotation and integration framework used to process heterogeneous datasets by exploiting semantic knowledge and improve the query processing in the large scale infrastructure. It is based on coarse-grained and fine-grained annotation models. The coarse-grained annotation employs Yago and SEeds SEarch to categorize domain information on big data and fine-grained annotation to enable semantic enrichment, integrate structured and unstructured data to form a global resource description framework ontology. With respect to cited approaches, our proposed model tackles the integration problem at a different and more abstract level. In particular, our approach is based on several abstractions for representing data, i.e. it makes use of both semantic-linguistic and multimedia features, with the aims of elevating raw data to information and finally, knowledge. Moreover, the implementation of such a model as a NoSQL general graph database has valuable benefits in the knowledge sharing process. Given that entities (e.g. concepts) stored in our knowledge base are closer to human understanding, a proper mapping of data to be ingested is required through entity pre-processing and disambiguation steps.

Peer2Peer technology has wildly used to share knowledge and we focus our attention on ontology based systems. In the last years several approaches and systems have been presented to manage and share knowledge and in particular ontologies. In [2] is presented InfoQuilt, a system for sharing ontologies in a peer-to-peer environment. Using this system a user can find relevant sets of ontologies, reuse them, create new ones and advertise the resulting ontologies. The system allows to search concepts and services exploring inter-ontological relationships. The implemented system is based on agents that allow users to request information, semantically correlate data from different sources and of heterogeneous type or representation; they can have an interactive interface for knowledge discovery. Becker et al. [5] propose a P2P extension for ontology editing based on Ontorama [26]. The system uses a sharing protocol using RDF. This approach provides a novel editing environment compared to the classic client/server ontology management approaches. In [3] a method to integrate different knowledge sources as thesaurus, gazetteer and a chronology based in an ontology using Topic Maps is proposed. This ontology is shown to Government Agencies by Web Services to support information harmonization about environmental data. KAON [25] is an open-source software infrastructure to manage ontologies for semantic-driven applications. It integrates traditional technologies as relational databases with new knowledge representation tools. There are several additional components which increase the KAON functionalities. COE (Cooperative Ontology Editor) [28] is a P2P application designed to allow ontology developers to share their knowledge through many activities: ontology sharing, ontology reuse and other traditional peer-to-peer mechanisms. It is implemented over COPPEER, a framework for creating flexible collaborative P2P applications that provides non-specific collaboration tools as plug-ins. SWAP (Semantic Web and Peer-to-Peer) [37] is a project that allows participants to keep private knowledge structures in their personal computer and share that knowledge in a P2P architecture. Users can extract ontologies from selected remote repositories, which are automatically integrated in their local repository. Any change in the source of the information is propagated to the local repositories. Oyster [44] is a java-based system, which assists users in managing, searching and sharing ontology metadata in a peer-to-peer network; it is a P2P application that exploits semantic web techniques in order to provide a solution for exchanging and re-using ontologies. The Oyster client on its own (e.g. disconnected from the P2P network) provides added value to it's users as it will give researchers an overview and search facilities of his/her own ontology metadata. In order to provide this functionalities, Oyster implements a proposal for a metadata standard, so called Ontology Metadata Vocabulary (OMV) which is based on discussions and agreement in the EU IST thematic network of excellence Knowledge Web as the way to describe ontologies. In [49] different techniques and tools for ontology definition and management are proposed together with a model for representing knowledge and a system based on peer-to-peer (P2P) paradigm to share general and domain knowledge. In [42] the authors propose an integrated agent system for ontology sharing on WWW, which enables users to manage ontologies and Semantic Web Services. The proposed system has several modules to manage personal information, translate them into standard language as RDF and analyze RDF to obtain user's interests and create Semantic Web Services which enable agent program to make inferences from grounding data on personalized ontology. In this context we notice also XAROP [11], a P2P platform to knowledge management in a decentralized IT infrastructure. Several surveys and books have been presented in the last years to evidence the importance of Peer-to-Peer and ontologies for enabling the Semantic Web; an useful reference is [52].

Even if the discussed literature tackles the problem of big data *Variety* using semantics and ontologies at different levels, the proposed models and tools are often limited to specific tasks and applications related to very specific domains or they do not consider multimedia data.

In our vision, a formal semantic-based model is needed to represent in a whole information about specific and general knowledge domains and the use of standard multimedia features is a basic step in order to overcome the long-standing issue of heterogeneity. Moreover, the use of a formal knowledge representation together with its implementation in a graph db allow the sharing of knowledge in a real scenario based on efficient techniques.

3   The Proposed Model

In this section, the proposed model to represent multimedia big data is presented with a description of its components and properties. Our model is based on property-based graph, which allows users to represent concepts and logical relations between them through a graph-based structure with an implicit agreement about the meaning of edges, nodes, labels and properties. The access to nodes and relationships in a native graph database is an efficient, constant-time operation and allows to quickly traverse millions of connections per second. Nodes are the entities in the graph. They can hold any number of attributes (key-value-pairs) called *properties*. Nodes can be tagged with *labels* representing their different roles in a knowledge domain. Relationships provide directed, named and semantically relevant connections between two node-entities. A relationship always has a direction, typology, start node and end node. Similar to nodes, relationships can also have properties. In most cases, relationships have quantitative properties, such as weights, costs, distances, ratings, time intervals, or strengths. Since our model is based on ontologies, this discussion starts with some notions about ontologies and the way to build them. Starting from some definitions of ontology [29,43] we extend them using also visual data to denote a concept; these data are represented using visual low-level features defined in MPEG-7 standard and others presented in literature described in the following. Thus an ontology can be seen as a set of "signs" and "relations" among them, denoting the concepts that are used in a knowledge domain.

The proposed model is composed of a triple $\langle S, P, C \rangle$ where:

S is a set of signs;
P is a set of properties used to link the signs in S;
C is a set of constraints on P.

In order to avoid confusion about the used terminology, we explicit point out that from the ontological point of view, the term property has a different meaning from the same term used in the property-based graph model; while in the first case, a property is a relation between two entities, in the second case it is intended as an attribute of an entity. In this context signs are words and visual data. The properties are linguistic, semantic and multimedia relations, and the constraints are validity rules applied to properties with respect to the multimedia category considered. In the proposed approach, knowledge is represented by an ontology implemented with respect to a semantic network (SN). A semantic network can be seen as a graph where the nodes are signs and arcs are relations between signs. A concept represents an abstract or general idea, something that is conceived in the human mind. It is the abstract representation of an object or a set of objects sharing some common features. In addition, an abstract concept is also represented by means of visual data, i.e. global and local feature vectors extracted from images depicting a given concept. The top-level ontological model is described in [50] using the *Web Ontology Language* (OWL). The two main classes in this meta-model are: *Concepts*, in which all objects are defined as individuals, and *Multimedia* (MM), which represents all the multimedia representations for Concepts, i.e. they represent all the "signs" in the ontology. From a linguistic-semantic point of view we exploit WordNet [39] to build the corresponding type of relations between signs. WordNet is a large lexical database of English. Nouns, verbs, adjectives and adverbs are grouped into sets of cognitive synonyms (synsets), each expressing a distinct concept. Synsets are interlinked by means of conceptual-semantic and lexical relations. The semantic and lexical properties are arranged in a hierarchy. The use of a linguistic approach allows an extension of linguistic properties also to multimedia data; e.g. different multimedia information related to the same concept are synonyms and in the same way hypernym/hyponym or meronym properties entail a semantic relation among the multimedia representation of concepts. Concepts, multimedia and properties are arranged in a class hierarchy resulting from the syntactic category for concepts and words, data type for multimedia and semantic or lexical for the properties. From a logical point of view, a multimedia representation can be related to all kind of concepts.

3.1   Ontological Model Formalization

In recent years, several languages have been proposed to represent ontologies and we choose to use OWL [24] due to its expressive power useful for our purposes and its extensive use in knowledge based systems. In our approach we use the DL version of OWL, because it is sufficiently effective to describe our model and its implementation. The DL version allows the declaration of disjoint classes, which may be used to assert that a word belongs to a syntactic category. Moreover, it allows the declaration of union classes used to specify domains and property ranges used to relate concepts and words belonging to different lexical categories. The ontology schema and the corresponding semantic network representation is formally described using OWL. Every node (both concept and multimedia) is an OWL individual. The connecting edges in the semantic network are represented as *ObjectProperties*. The considered linguistic properties for Concepts are shown in Table 1.

Table 1. Linguistic properties

Lexical properties	Synonym, antonym, pertainym, nominalization, derived from adjective, participle of verb
Semantic properties	Hypernyms, hyponyms, coordinate terms, holonym, meronym, hypernym, troponym, entailment, related nouns, similar to, coordinate terms, Participle of verb, root adjectives

These properties have constraints that depend on the syntactic category (noun, verb, adjective, adverb) or kind of semantic or lexical properties. For example, the hyponymy property can only relate nouns to nouns or verbs to verbs. A semantic property may links concepts to concepts respecting the constraints defined by the ontological model. Concept and multimedia are considered with *DatatypeProperties*, which relate individuals to pre-defined data types. Each multimedia is related to the concept it represents by the ObjectProperty *hasConcept*, whereas a concept is related to multimedia that represents it using the ObjectProperty *hasMM*. These are the only properties that can relate concepts to multimedia and vice versa; all of the other properties relate multimedia to multimedia and concepts to concepts. The two main classes are not supposed to have common elements; therefore they are defined as disjoint. The class MM defines the logical model of the multimedia forms used to express a concept. On the other hand, the class Concept represents the meaning related to a multimedia form; the sub-classes have been derived from related categories. There are some union classes that are useful for defining the properties of domain and co-domain. Attributes have been defined for Concept and MM respectively; Concept has: *Name* that represents the concept name; *Description* that gives a short description of concept. On the other hand MM has *Name* as attribute that is the MM name and a set of features described in Table 2. More details and references to such descriptors are reported in Sect. 3.2.

All elements have an *ID* within a unique identification number. Table 3 shows some of the properties considered and their domains and ranges of definition.

The use of domain and codomain reduces the property range application; however, the model as described so far does not exhibit perfect behavior in some

Table 2. Visual features

Data type	Features
Visual	Auto Color Correlogram (ACC), Scalable Color (SC) Fuzzy Color and Texture Histogram (FCTH), Color Layout (CL), Edge Histogram (EH), Color and Edge Directivity Descriptor (CEDD), Joint-Composite Descriptor (JCD), Pyramid Histogram of Oriented Gradients (PHOG)

Table 3. Property features

Property	Domain	Range
hasMM	Concept	MM
hasConcept	MM	Concept
hypernym	NounsAnd VerbsConcept	NounsAnd VerbsConcept
holonym	NounConcept	NounConcept
entailment	VerbConcept	VerbConcept
similar	AdjectiveConcept	AdjectiveConcept

cases. For example, the model does not know that a hyponymy property defined on sets of nouns and verbs would have 1) a range of nouns when applied to a set of nouns and 2) a range of verbs when applied to a set of verbs. Therefore, it is necessary to define several *constraints* to express the ways that the linguistic properties are used to relate concepts and/or MM. Table 4 shows some of the defined constraints specifying the classes to which they have been applied with respect to the properties considered. The table also shows the matching range.

Table 4. Model constraints

Costraint	Class	Property	Constraint range
AllValuesFrom	NounConcept	hyponym	NounConcept
AllValuesFrom	AdjectiveConcept	attribute	NounConcept
AllValuesFrom	NounWord	synonym	NounWord
AllValuesFrom	VerbWord	also_see	VerbWord

Sometimes, the existence of a property between two or more individuals entails the existence of other properties. For example, since the concept "dog" is a hyponym of "animal", animal is a hypernym of dog. These characteristics are represented in OWL by means of property features. Table 5 shows several of those properties and their features.

Table 5. Property features

Property	Features
hasMM	*inverse* of hasConcept
hasConcept	*inverse* of hasMM
hyponym	*inverse* of hypernym; *transitivity*
hypernym	*inverse* of hyponym; *transitivity*
cause	*transitivity*
verbGroup	*symmetry* and *transitivity*

The proposed model allows a high-level conceptual matching using different type of low-level representations. Moreover, an ontology built using this model can be used to infer information by means of formal representation of properties among multimedia data and concepts.

3.2 Property-Based Graph Model Formalization

We propose the use of a semantic approach to formalize the model structure of multimedia BigData. The use of multimodal features to represent concepts and linguistic properties to relate them are an effective way to bridge the gap between the target semantic classes and the available low-level multimedia descriptors. The proposed model has been implemented in a NoSQL graphdb populated from different knowledge sources and a visualization of this very large knowledge base has been discussed as a case study. We decided to store as nodes in the graph database model the two main classes of the ontological model, i.e. Concepts and Multimedia. Therefore the nodes in the graph may have two different labels, *Concept* and *Multimedia* and a different number of properties(attributes) according to it.

The nodes labelled with concept label have the following attributes:

- *id*, a number used to identify each concept in the database.
- *sid*, the id used in WordNet for that concept.
- *pos*, the part of speech of the concept. It can be a noun, a verb, an adjective or an adverb.
- *lemmas*, the list of english terms used to represent the concept.
- *glossary*, a brief description of the concept, explaining its meaning.

Multimedia nodes contain visual information. In particular, global descriptors have been extracted from the images considered in our knowledge base. Global descriptors are feature vectors extracted from images considering them as a whole unit. In particular, the nodes with the label multimedia share the following attributes:

- *id*, a number used to identify each multimedia in the database.
- *url*, the path to the multimedia file representing the concept.

- *PHOG* [9], the extracted Pyramid Histogram of Oriented Gradients (PHOG) feature vector.
- *JCD* [20], the extracted Joint-Composite Descriptor(JCD) feature vector.
- *CEDD* [21], the extracted Color and Edge Directivity Descriptor (CEDD) feature vector.
- *SC* [36], the extracted Scalable Color(SC) feature vector.
- *EH* [54], the extracted Edge Histogram (EH) feature vector.
- *FCTH* [22], the path to the Fuzzy Color and Texture Histogram (FCTH) feature vector.
- *CL* [31], the path to the Color Layout(CL) feature vector.
- *ACC* [53], the path to the Auto Color Correlogram(ACC) feature vector.

Also relationships have some attributes:

- *id*, univoque identifier of relation in the graph database
- *type*, the type of relation, i.e. hyponym, antonym, hasConcept, etc.
- *weight*, a number in the interval $[0, 1]$ used to assign a strength level to each type of relation and consequently giving the possibility to perform similarity metrics based on weighted distances.

Relationships can link multimedia nodes to multimedia nodes, concept nodes to concept nodes, multimedia nodes to concept nodes and vice-versa. The presence of a weight attribute is desirable and needed for our model, since it allows to discriminate between different relationships, which is the case of a semantic-based model where many relationships are present. Such weights may be actively exploited during analysis and usage of metrics for sophisticated path distances calculations. Using a set-theory notation, we give here formal definitions for them. Γ is the entire network containing nodes and arcs, V is the set of all nodes, E is the set containing all edges, C is the set of concept nodes, and M is the set of multimedia nodes. From a theoretical point of view, relationships are formally defined as hyperarcs in a hypergraph structure. The reason for this choice is that the hyperarc concept allows for more powerful and generalized definitions of "many-to-many" relationships between two sets of nodes.

Definition 1 *(Multimedia to Multimedia relationship M-M):*
Let $\hat{M} \subset M \subset V$ and $\mathring{M} \subset M \subset V$, with $\hat{M} \cap \mathring{M} = \varnothing$, two disjoint subsets of nodes of M in Γ, a multimedia to multimedia relationship is a weighted hyperarc $e_i \subset E$ with a weight $w \longrightarrow [0, 1]$ connecting the nodes in subset $\hat{M}$ with nodes in subset $\mathring{M}$.

Multimedia to Multimedia relationships are used for example to relate multimedia contents by exploiting metadata, features extracted from the contents, low-level multimedia descriptors, etc.

Definition 2 *(Concept to Multimedia relationship C-M):*
Let $\hat{C} \subset C \subset V$ and $\hat{M} \subset M \subset V$ two subsets of C and M respectively, in Γ, a concept to multimedia relationship is a weighted hyperarc $e_i \subset E$ with a weight $w \longrightarrow [0, 1]$ connecting the nodes in $\hat{C}$ with nodes in $\hat{M}$.

This relationship is called *hasMM* and it is used to define a link between the Concept and Multimedia nodes.

Definition 3 *(Multimedia to Concept relationship M-C):*
Let $\hat{M} \subset M \subset V$ and $\hat{C} \subset C \subset V$ two subsets of nodes of M and C respectively, in Γ, a multimedia to concept relationship is a weighted hyperarc $e_i \subset E$ with a weight $w \longrightarrow [0,1]$ connecting the nodes in $\hat{M}$ with nodes in $\overset{\circ}{C}$.

With this relationship we are able to associate a multimedia "sign" to a set of concepts. As previously described, we use the *hasConcept* ObjectProperty defined in the top level ontological model. In this formalization, each multimedia is related to the concept it represents by the *hasConcept*, whereas a concept is related to multimedia that represent it using *hasMM*.

Definition 4 *(Concept to Concept relationship C-C):*
Let $\hat{C} \subset C \subset V$ and $\overset{\circ}{C} \subset C \subset V$, with $\hat{C} \cap \overset{\circ}{C} = \varnothing$, two disjoint subsets of C in Γ, a concept to concept relationship is a weighted hyperarc $e_i \subset E$ with a weight $w \longrightarrow [0,1]$ connecting the nodes in $\hat{C}$ with nodes in $\overset{\circ}{C}$.

This kind of link is used to exploit the semantic and linguistic properties between Concept nodes, described in Sect. 3.1.

The use of general top level ontological model for Multimedia and Concepts allows us to exploit all the potentials of ontologies, highlighting the importance of a strong formalization and organization of data. Moreover, the linguistic properties used to relate concepts give a formalization of our representation closer to human languages.

3.3   Semantic Multimedia Big Data Population

In this section we provide details related to the implementation and population of the knowledge graph. The general architecture of the system is shown in Fig. 1.

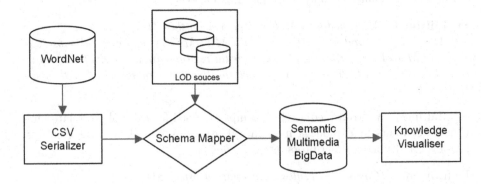

Fig. 1. System architecture

The main component of the architecture is the *Semantic Multimedia Big-Data* (SMBD) which contain the implemented model. In particular, the model previously described has been implemented as instance of a Neo4J [40] graph database, according to the property-based labelled graph model. Neo4J is a NoSQL graph database entirely written in Java language. It guarantees reliability for transactions by means of *ACID* (i.e. Atomicity, Consistency, Integrity, Durability) properties. It is possible to query the database through a powerful language, named *Cypher*. We consider concepts and multimedia representations as graph nodes, whereas semantic, linguistic, semantic-linguistic and multimedia relations as edges connecting nodes. For example, the hyponymy property is converted in an edge that links two concept nodes (nouns to nouns or verbs to verbs). The database was built by means of LOAD Cypher queries of information stored in CSV files. This method allows to dramatically increase the speed during the database creation phase. First, we created the CSV files (CSV Serializer block) containing nodes and relationships exploiting the Java library *JWNL*, the dictionary *WordNet 3.0*, images related to the concepts were collected from ImageNet and BabelNet and feature extracted with the library LIRE. Then, we loaded in Neo4J nodes and relationships. Linked Open Data sources can be ingested as well into the database after that a proper mapping with concepts in the graph database is found ny the *Schema Mapper* module. The Knowledge Visualiser is the module used for the visualization task. In particular, a well-known technique based on the library D3.js [56] has been used for enhancing the Neo4J browser, which is the standard tool used for visualizing data stored in the graph. In Table 6 we report some statistics related to our instance of the database used as a case study.

Table 6. Database statistics

Type of entity	#
Concept nodes	117659
Multimedia nodes	302305
Concept nodes properties	675401
Multimedia nodes properties	1776195
Linguistic-semantic relationships	377585
Cross-label relationships	605185

4   A Case Study for Sharing Knowledge

We provide an example of knowledge representation and sharing showing the power of combining both textual and visual information in a peer-to-peer environment.

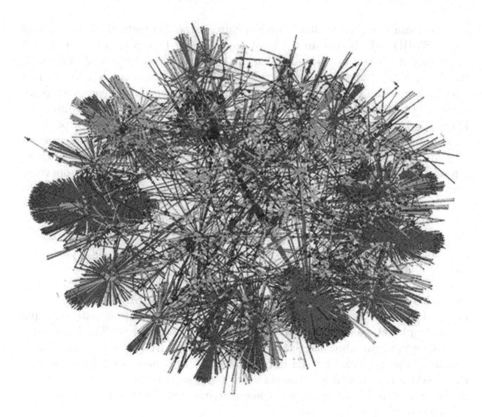

Fig. 2. Partial visualization of our graph

In order to give an idea to the reader about complexity of our network we show in Fig. 2 a partial view of it containing 8000 relationships and 5990 nodes. It has been obtained using the *Apoc* plugin and *Gephi* an open-source software for visualization and exploration of graphs and networks. The layout used for this representation is the *Force Atlas 2*, with the *dissolve Hubs* and *prevent overlap* settings.

The comparison between the numbers related to Fig. 2 and the statistics of our knowledge base provided in Table 6 help to understand the huge size and complexity of the network. Figure 3 gives a scratch of our database in order to visualize the structure of the logical model in a more meaningful way. It is the result of a Cypher query to highlight the *hasMM* relationship between nodes labelled as a Concept (light blue nodes) and nodes labelled as a Multimedia (green nodes).

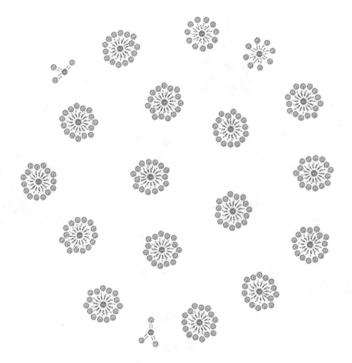

Fig. 3. Excerpt of graph database (Colour figure online)

We consider as case study a knowledge domain related to famous painters. As a matter of fact, despite WordNet is usually intended only as a dictionary or as a thesaurus containing general concepts representing classes, in its newer versions it also contains many individuals or instances, linked to classical Word-Net concepts by means of different relationships, named *instance_hyponym* and its inverse relation *instance_hypernym*. This is the case, for example, of famous painters. One of the synsets contained in WordNet, named *old_master*, describes famous painter of the past, also providing the following definition: "a great European painter prior to 19th century". Therefore we can extract all the instances for the Concept representing "old masters" querying our Knowledge Base.

The proposed model has been implemented in our graph db used to supply knowledge system to create in a peer to peer environment.

The system has this common model for defining knowledge structure and contents. In this way all peers have a common view of distributed knowledge and they can share it in a simple way. This structure is implemented in a Semantic Multimedia Bigdata based on a graph DBMS.

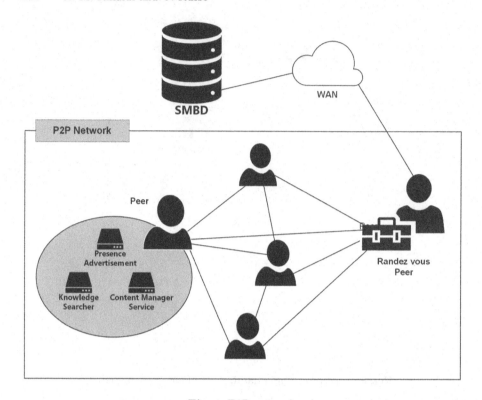

Fig. 4. P2P network

The proposed system has several software modules and, from a top-level view, they can be organized around some entities and macro-functionalities. The main system entities are: *Peer:* it is the agent in charge of editing and managing local knowledge (i.e. graphs extracted from the SMDB); each user which takes part in the network is a peer; *Rendez-Vous Peer:* its task is to build the P2P network, manage a list of sharing local knowledge between peers and allow the communication with the SMDB. The general architecture of the system is hereinafter described and it is drawn in Figure 4 together with an example of each single macro-module.

A peer has two main tasks: (i) managing and editing extracted graphs and (ii) putting in share local graphs. A Rendez-Vous peer has a list of active peers and a description of their contents. It uses these information in the knowledge discovery step between peers.

In each single peer a system interface shows the catalog of the graphs stored in the local repository to the user by means of an appropriate software module called OntoSearcher; OntoSearcher performs a syntactic search or a browsing in a directory structure arranged by arguments with the aim of finding a graph relevant to the user interest. When OntoSearcher finds a suitable graph, the OntoViewer builds a view to represent it. A user can modify the graph, add contents or build a new one with the peer editing functionalities. On the other hand a peer must communicate to the other peer and with the Rendez-Vous one for sharing ontologies. JXTA is the framework used to build the P2P network; it uses advertisements in the communication steps. In the following subsections are described into details both the remaining modules drawn in Fig. 4.

Many information systems use a knowledge base to represent data in order to satisfy information requests and in the author's vision it is a good choice for having a common view of the same general and specific knowledge domain.

In the proposed framework we use our SMBD as a "starting point" for users because they can extract an initial graph from this general knowledge base and expand it to have a specialized one; these tasks are explained in the following of this section. Hence, efficiency and performances of the knowledge sharing process depend on user actions and on the JXTA framework performances.

The graph is built using an ad-hoc interface based on cypher language.

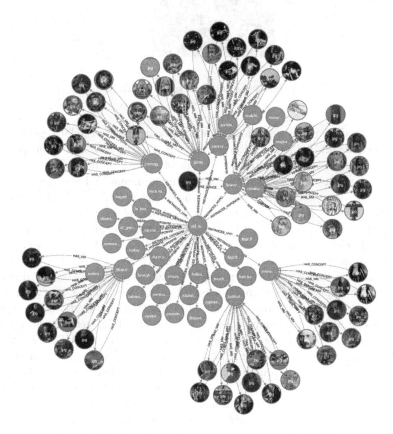

Fig. 5. Visualization of query result obtained with D3

Fig. 6. Data exploration of query result with D3

Using the OntoEditor functionalities a user can modify the local graph structure as a whole adding new MM and Concepts, linking MM and Concepts using arrows (lexical and semantic properties), deleting nodes and arcs.

The images are fetched using a search engine image tool (i.e. google image) by means of a query with the concept name in the SMBD or added by the user. In addition, the user can use words from the SMBD concept description or other ones manually added to refine his/her search. Once images have been fetched, they can be added to the considered concept using an ad hoc interface.

At this step of our research we are interested in showing a real implementation of our model. Therefore, the proposed methods and techniques are described and tested by means of a complete use case in order to put in evidence the several features of the proposed model and implemented system. The system has been completely developed using Java and the P2P network is based on JXTA libraries. The process of extracting a graph from the SMBD begins with an interaction in which the user inserts a specific term by means of the user interface and chooses the proper sense by reading the description of the related concepts. The system retrieves the correct sense and builds the graph following the steps described in the previous section. An example of results is shown in Fig. 5 where a graph regarding *old master* concept is drawn together with some properties among the related concepts; moreover a user can interact with the system editing tools.

The images are displayed with a circular shape in the multimedia nodes by using the D3 library. The user can also click on the node he/she is interested in to visualize the image in its original dimension. As an example, Fig. 6 shows an image related to the famous painter *Raffaello Sanzio.*

5   Conclusion and Future Works

The constant production of digital data coming from different sources gives us tons of data, from which we could extract precious and useful information. This issue becomes more interesting if we consider multimedia data. In this context, the use of a formal model to represent and manage data is a silver bullet task to implement intelligent information systems. The aim of our work has been to provide a novel model to represent in a formal and complete way the knowledge structure implemented in a generalized semantic multimedia big data. In this paper we have described the problem of data heterogeneity and the impact of multimedia data. In this scenario, we propose a formal model combining top-level ontology models and graph models represented by a labelled, property-based structure to take into account both semantic, linguistic and multimedia aspects. The use of a common model and it's implementation has been used in a P2P environment to share knowledge. A complete case study in a given scenario shows the real use of our model and its expressive power. The distinctive features of our framework is to use a simple and general formal model for multimedia knowledge representation taking into account a linguistic approach considered as the natural communication way between human agents integrating standard descriptions for multimedia data. The proposed model and its implementation can be used in crucial knowledge based applications for mining information. In this first stage of our research, some aspects related to efficiency and scalability with big volumes of data are not taken into account. Our goal is in describing our semantic-based model together with a first prototype implementation. The current research effort is based on the use of the proposed model together with multimedia similarity metrics for content based analysis and we count to improve the exploitation of multimedia data enriching our knowledge base and developing new classification algorithms based on it. There are other research lines to be investigated as the implementation of a system based on our model in different applications domains with a particular interest to multimodal information retrieval [46], the implementation of other multimedia representation as audio features and the definition of strategies to integrate other heterogeneous knowledge sources [51]. In addition, we are also interested in finding more efficient techniques to visualize and analyze data stored in very large knowledge bases [14,18] and to define strategies and perform experiments for the evaluation of our model and approaches through statistical and quantitative measures, as well as to test the efficiency and performances of the knowledge sharing process. Moreover, we plan to extend the proposed approach for sharing knowledge among different kind of agents both software and robotics by designing a Knowledge-as-a-Service (KaaS) architecture.

References

1. Agerri, R., Artola, X., Beloki, Z., Rigau, G., Soroa, A.: Big data for natural language processing: a streaming approach. Knowl.-Based Syst. **79**, 36–42 (2015)
2. Arumugam, M., Sheth, A.P., Arpinar, I.B.: Towards P2P semantic web: a distributed environment for sharing semantic knowledge on the web. In: Proceedings of the Workshop on Real World RDF and Semantic Web Applications (2002)
3. Bandholtz, T.: Sharing ontology by web services: Implementation of a semantic network service (SNS) in the context of the German environmental information network (Gein). In: Proceedings of SWDB 2003, pp. 189–201 (2003)
4. Bansal, S.K., Kagemann, S.: Integrating big data: a semantic extract-transform-load framework. Computer **48**(3), 42–50 (2015)
5. Becker, P., Eklund, P., Roberts, N.: Peer-to-peer based ontology editing. In: Proceedings of NWESP 2005, p. 259. IEEE Computer Society, Washington (2005). https://doi.org/10.1109/NWESP.2005.63
6. Bello-Orgaz, G., Jung, J.J., Camacho, D.: Social big data: recent achievements and new challenges. Inf. Fusion **28**, 45–59 (2016)
7. Benbernou, S., Huang, X., Ouziri, M.: Semantic-based and entity-resolution fusion to enhance quality of big RDF data. IEEE Trans. Big Data, (Early Access), 1 (2017). https://ieeexplore.ieee.org/document/7937830
8. Berners-Lee, T., Hendler, J., Lassila, O.: The semantic web. Sci. Am. **284**(5), 34–43 (2001)
9. Bosch, A., Zisserman, A., Munoz, X.: Representing shape with a spatial pyramid kernel. In: Proceedings of the 6th ACM International Conference on Image and Video Retrieval, pp. 401–408 (2007)
10. Boury-Brisset, A.C.: Managing semantic big data for intelligence. In: STIDS, pp. 41–47 (2013)
11. Tempich, C.: XAROP: a midterm report in introducing a decentralized semantics-based knowledge sharing application. In: Karagiannis, D., Reimer, U. (eds.) PAKM 2004. LNCS (LNAI), vol. 3336, pp. 259–270. Springer, Heidelberg (2004). https://doi.org/10.1007/978-3-540-30545-3_25
12. Caldarola, E., Rinaldi, A.: A multi-strategy approach for ontology reuse through matching and integration techniques. Adv. Intell. Syst. Comput. **561**, 63–90 (2018). https://doi.org/10.1007/978-3-319-56157-8_4
13. Caldarola, E.G., Picariello, A., Rinaldi, A.M.: An approach to ontology integration for ontology reuse in knowledge based digital ecosystems. In: Proceedings of the 7th International Conference on Management of Computational and Collective intElligence in Digital EcoSystems, pp. 1–8. ACM (2015)
14. Caldarola, E.G., Picariello, A., Rinaldi, A.M.: Big graph-based data visualization experiences: The wordnet case study. In: 2015 7th International Joint Conference on Knowledge Discovery, Knowledge Engineering and Knowledge Management (IC3K), vol. 1, pp. 104–115. IEEE (2015)
15. Caldarola, E.G., Picariello, A., Rinaldi, A.M.: Experiences in wordnet visualization with labeled graph databases. Commun. Comput. Inf. Sci. **631**, 80–99 (2016)
16. Caldarola, E.G., Rinaldi, A.M.: Big data: a survey. In: Proceedings of 4th International Conference on Data Management Technologies and Applications, pp. 362–370. SCITEPRESS-Science and Technology Publications, Lda (2015)
17. Caldarola, E.G., Rinaldi, A.M.: An approach to ontology integration for ontology reuse. In: 2016 IEEE 17th International Conference on Information Reuse and Integration (IRI), pp. 384–393. IEEE (2016)

18. Caldarola, E.G., Rinaldi, A.M.: Big data visualization tools: a survey: the new paradigms, methodologies and tools for large data sets visualization. In: Proceedings of the 6th International Conference on Data Science, Technology and Applications-DATA 2017, pp. 296–305 (2017)
19. Caldarola, E.G., Rinaldi, A.M.: Modelling multimedia social networks using semantically labelled graphs. 2017 IEEE International Conference on Information Reuse and Integration (IRI), pp. 493–500 (2017)
20. Chatzichristofis, S., Boutalis, Y., Lux, M.: Selection of the proper compact composite descriptor for improving content based image retrieval. In: Proceedings of the 6th IASTED International Conference, vol. 134643, p. 064 (2009)
21. Chatzichristofis, S.A., Boutalis, Y.S.: CEDD: color and edge directivity descriptor: a compact descriptor for image indexing and retrieval. In: Gasteratos, A., Vincze, M., Tsotsos, J.K. (eds.) ICVS 2008. LNCS, vol. 5008, pp. 312–322. Springer, Heidelberg (2008). https://doi.org/10.1007/978-3-540-79547-6_30
22. Chatzichristofis, S.A., Boutalis, Y.S.: FCTH: fuzzy color and texture histogram-a low level feature for accurate image retrieval. In: 2008 Ninth International Workshop on Image Analysis for Multimedia Interactive Services, pp. 191–196. IEEE (2008)
23. De Mauro, A., Greco, M., Grimaldi, M.: A formal definition of big data based on its essential features. Libr. Rev. 65(3), 122–135 (2016)
24. Dean, M., Schreiber, G.: OWL web ontology language reference. Technical report, W3C, February 2004. http://www.w3.org/TR/2004/REC-owl-ref-20040210/
25. Bozsak, E., et al.: KAON—towards a large scale semantic web. In: Bauknecht, K., Tjoa, A.M., Quirchmayr, G. (eds.) EC-Web 2002. LNCS, vol. 2455, pp. 304–313. Springer, Heidelberg (2002). https://doi.org/10.1007/3-540-45705-4_32
26. Eklund, P., Roberts, N., Green, S.: OntoRama: browsing RDF ontologies using a hyperbolic-style browser. In: Proceedings of CW 2002, p. 0405. IEEE Computer Society, Washington (2002)
27. Emani, C.K., Cullot, N., Nicolle, C.: Understandable big data: a survey. Comput. Sci. Rev. 17, 70–81 (2015)
28. Xexeo, G., et al.: Peer-to-peer collaborative editing of ontologies. In: Proceedings of CSCWD 2004, pp. 186–190 (2004)
29. Gruber, T.R.: A translation approach to portable ontology specifications. Knowl. Acquis. 5(2), 199–220 (1993)
30. Hassan, T., Cruz, C., Bertaux, A.: Ontology-based approach for unsupervised and adaptive focused crawling. In: Proceedings of The International Workshop on Semantic Big Data, p. 2. ACM (2017)
31. Kasutani, E., Yamada, A.: The MPEG-7 color layout descriptor: a compact image feature description for high-speed image/video segment retrieval. In: Proceedings 2001 International Conference on Image Processing (Cat. No. 01CH37205), vol. 1, pp. 674–677. IEEE (2001)
32. Knoblock, C.A., Szekely, P.: Exploiting semantics for big data integration. AI Mag. 36(1), 25–38 (2015)
33. Lee, S., Chinthavali, S., Duan, S., Shankar, M.: Utilizing semantic big data for realizing a national-scale infrastructure vulnerability analysis system. In: Proceedings of the International Workshop on Semantic Big Data, p. 3. ACM (2016)
34. Lv, Z., Song, H., Basanta-Val, P., Steed, A., Jo, M.: Next-generation big data analytics: state of the art, challenges, and future research topics. IEEE Trans. Ind. Inform. 13(4), 1891–1899 (2017)

35. Mami, M.N., Scerri, S., Auer, S., Vidal, M.-E.: Towards semantification of big data technology. In: Madria, S., Hara, T. (eds.) DaWaK 2016. LNCS, vol. 9829, pp. 376–390. Springer, Cham (2016). https://doi.org/10.1007/978-3-319-43946-4_25
36. Manjunath, B.S., Ohm, J.R., Vasudevan, V.V., Yamada, A.: Color and texture descriptors. IEEE Trans. Circuits Syst. Video Technol. **11**(6), 703–715 (2001)
37. Ehrig, M., et al.: SWAP: ontology-based knowledge management with peer-to-peer. In: Izquierdo, E. (ed.) Proceedings of WIAMIS 2003, pp. 557–562. World Scientific, London (2003)
38. Mezghani, E., Exposito, E., Drira, K., Da Silveira, M., Pruski, C.: A semantic big data platform for integrating heterogeneous wearable data in healthcare. J. Med. Syst. **39**(12), 185 (2015)
39. Miller, G.A.: WordNet: a lexical database for English. Commun. ACM **38**(11), 39–41 (1995)
40. Miller, J.J.: Graph database applications and concepts with Neo4j. In: Proceedings of the Southern Association for Information Systems Conference, Atlanta, GA, USA, vol. 2324, p. 36 (2013)
41. Moscato, V., Picariello, A., Rinaldi, A.M.: A recommendation strategy based on user behavior in digital ecosystems. In: Proceedings of the International Conference on Management of Emergent Digital EcoSystems, pp. 25–32. ACM (2010)
42. Nakayama, K., Hara, T., Nishio, S.: An agent system for ontology sharing on WWW. In: Proceedings of WWW 2005, pp. 964–965. ACM, New York (2005)
43. Neches, R., et al.: Enabling technology for knowledge sharing. AI Mag. **12**(3), 36–56 (1991)
44. Palma, R., Haase, P., Gómez-Pérez, A.: Oyster: sharing and re-using ontologies in a peer-to-peer community. In: Proceedings of WWW 2006, pp. 1009–1010 (2006)
45. Purificato, E., Rinaldi, A.M.: Multimedia and geographic data integration for cultural heritage information retrieval. Multimedia Tools Appl. **77**(20), 27447–27469 (2018). https://doi.org/10.1007/s11042-018-5931-7
46. Purificato, E., Rinaldi, A.M.: A multimodal approach for cultural heritage information retrieval. In: Gervasi, O., et al. (eds.) ICCSA 2018. LNCS, vol. 10960, pp. 214–230. Springer, Cham (2018). https://doi.org/10.1007/978-3-319-95162-1_15
47. Quboa, Q., Mehandjiev, N.: Creating intelligent business systems by utilising big data and semantics. In: 2017 IEEE 19th Conference on Business Informatics (CBI), vol. 2, pp. 39–46. IEEE (2017)
48. Rani, P.S., Suresh, R.M., Sethukarasi, R.: Multi-level semantic annotation and unified data integration using semantic web ontology in big data processing. Cluster Comput. **22**(5), 10401–10413 (2017). https://doi.org/10.1007/s10586-017-1029-7
49. Rinaldi, A.: A peer-to-peer system to share ontology in the semantic web. In: Proceedings of the 5th International Conference on Soft Computing as Transdisciplinary Science and Technology, CSTST 2008 - Proceedings, pp. 644–649 (2008)
50. Rinaldi, A.M.: A multimedia ontology model based on linguistic properties and audio-visual features. Inf. Sci. **277**, 234–246 (2014)
51. Rinaldi, A.M., Russo, C.: A matching framework for multimedia data integration using semantics and ontologies. In: 2018 IEEE 12th International Conference on Semantic Computing (ICSC), pp. 363–368. IEEE (2018)
52. Staab, S., Stuckenschmidt, H.: Semantic Web and Peer-to-Peer - Decentralized Management and Exchange of Knowledge and Information, 1st edn., p. 365. Springer, Heidelberg (2006). https://doi.org/10.1007/3-540-28347-1
53. Tungkasthan, A., Intarasema, S., Premchaiswadi, W.: Spatial color indexing using ACC algorithm. In: 2009 7th International Conference on ICT and Knowledge Engineering, pp. 113–117. IEEE (2009)

54. Won, C.S., Park, D.K., Park, S.J.: Efficient use of mpeg-7 edge histogram descriptor. ETRI J. **24**(1), 23–30 (2002)
55. Xu, Z., et al.: Knowle: a semantic link network based system for organizing large scale online news events. Future Gener. Comput. Syst. **43**, 40–50 (2015)
56. Zhu, N.Q.: Data Visualization with D3.js Cookbook. Packt Publishing Ltd., Birmingham (2013)
57. Zomaya, A.Y., Sakr, S.: Handbook of Big Data Technologies. Springer, Heidelberg (2017). https://doi.org/10.1007/978-3-319-49340-4

Facilitating and Managing Machine Learning and Data Analysis Tasks in Big Data Environments Using Web and Microservice Technologies

Shadi Shahoud$^{(\boxtimes)}$, Sonja Gunnarsdottir, Hatem Khalloof,
Clemens Duepmeier, and Veit Hagenmeyer

Institute for Automation and Applied Informatics (IAI),
Karlsruhe Institute of Technology (KIT), Karlsruhe, Germany
{shadi.shahoud,hatem.khalloof,clemens.duepmeier,veit.hagenmeyer}@kit.edu,
sonjabara@gmail.com

Abstract. Driven by the current advances of machine learning in a wide range of application areas, the need for developing easy to use frameworks for instrumenting machine learning effectively for non data analytics experts as well as novices increased dramatically. Furthermore, building machine learning models in the context of Big Data environments still represents a great challenge. In the present article, those challenges are addressed by introducing a new generic framework for efficiently facilitating the training, testing, managing, storing and retrieving of machine learning models in the context of Big Data. The framework makes use of a powerful Big Data software stack platform, web technologies and a microservice architecture for a fully manageable and highly scalable solution. A highly configurable user interface hiding platform details from the user is introduced giving the user the ability to easily train, test and manage machine learning models. Moreover, the framework automatically indexes and characterizes models and allows flexible exploration of them in the visual interface. The performance and usability of the new framework is evaluated on state-of-the-arts machine learning algorithms: it is shown that executing, storing and retrieving machine learning models via the framework results in a well acceptable low overhead demonstrating that the framework can provide an efficient approach for facilitating machine learning in Big Data environments. It is also evaluated, how configuration options (e.g. caching of RDDs in Apache Spark) affect runtime performance. Furthermore, the evaluation provides indicators for when the utilization of distributed computing (i.e. parallel computation) based on Apache Spark on a cluster outperforms single computer execution of a machine learning model.

Keywords: Microservice · Web-based applications · Big Data · Data analytic · Machine Learning

© Springer-Verlag GmbH Germany, part of Springer Nature 2020
A. Hameurlain et al. (Eds.) TLDKS XLV, LNCS 12390, pp. 132–171, 2020.
https://doi.org/10.1007/978-3-662-62308-4_6

1 Introduction

Data mining is the extraction of implicit, unknown and potentially useful information from data [35]. To this aim, Machine Learning (ML) provides the technical basis including algorithms, metrics and technologies. It is the process of taking an algorithm specification, providing training data and using a training procedure to learn model parameters that optimally fit the training data. The success of ML in many application areas such as text classification [4], speech recognition [5], medical diagnostics [6], energy generation forecasting [7] and load forecasting [8], to name a few, paved the road for more in-depth research on new methodologies as well as an even-growing demand for ready-to-go ML software solutions.

Although ML can be used for solving many complex business problems, there are also some downsides. Applying ML is usually a time-consuming process for the user, in which a lot of hyperparameters need to be configured to achieve the best performance in a so called trial-and-error approach. Such approach is based on the idea that all possible combinations of learning algorithms with their relevant parameters will be tried for each task until a good solution is found. However, this is typically inextricable. It wastes the resources for constructing multiple models which can take a long time especially in the case of large datasets to be forecasted.

Consequently and due to the rapid increase of data, more intelligent solutions utilizing Big Data platforms are becoming one of the hottest topics related to ML [9], where a distributed execution environment is required for the computation of larger datasets. Gaining insightful information, finding patterns and extracting knowledge from big datasets are quite complex tasks. Additionally, the configurations of the underlying Big Data infrastructure introduce more complexity for configuring and running ML tasks. This process consists of multiple steps and is commonly called Machine Learning Pipeline (MLP). Figure 1 shows a simplified MLP encompassing data preprocessing, splitting the data into training and test data, model training and model testing.

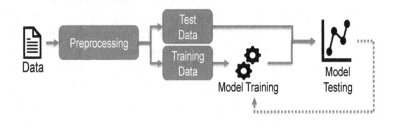

Fig. 1. Simplified methodology of Machine Learning Pipeline (MLP).

The aforementioned challenges are addressed in developing a new microservice-based solution by Shahoud et al. in [34]. They developed a new conceptual framework helping users to solve ML problems in Big Data environments without caring too much about technical issues of the underlying Big Data and cluster computing environment as runtime platform. The goal of this framework is to facilitate training, testing, managing, storing and retrieving ML models in the context of Big Data by using an easy to use web interface which hides the complexity of the underlying runtime environment from the user. For efficient scalable processing, the framework employs a Big Data cluster, a microservices architecture and modern web technologies like REST, React and Spring Boot. As a first exemplary application, smart grid applications are addressed in the evaluation. The proposed framework is able to perform ML tasks on energy time series datasets using a variety of algorithms on different types and size of such data.

In context of ML, the users can be categorized into two main categories, namely expert and non-expert ones as shown in Table 1. On one hand, the expert users have a deep understanding of ML and good programming skills to implement ML models using, for example, some developing tools like Jupiter Notebook[1]. They have worked with ML libraries before and are capable of programming algorithms themselves. On the other hand, non-expert users are grouped into two sub-categories. The first one includes the users who are familiar with statistics and ML but are not able to write the necessary script for training and evaluating ML models particularly in Big Data environments. This sub-category of users will be mainly supported by the current framework presented in this article. The second sub-category of non-expert users is inexperienced and not knowledgeable about statistics and ML. They need to have some analysis results using ML, but they only have the data and seems to be difficult for them to write or build ML models because they also do not have the required ML programming skills.

Table 1. User categories.

Category Nr.	User category		Properties
1	Expert		ML knowledge (+) ML Programming skills (+)
2	Non-expert	**A**	ML knowledge (+) ML Programming skills (−)
		B	ML knowledge (−) ML Programming skills (−)

For the evaluation of the basic concepts of the framework, a first implementation is developed which utilizes Apache Spark as runtime environment for ML on a Big Data cluster and spark.ml as a ML library [18]. The storage layer of the framework utilizes the Hadoop Distributed File System (HDFS) [10] and a PostgreSQL database [17] for storing the required input and the resulting output

---

[1] https://jupyter.org/.

data. To facilitate the building, training and running of ML models, an easy-to-use web User Interface (UI) which assists non-expert users in performing these tasks is conceptualized and implemented in the current version. The UI utilizes microservices [11] running on the Big Data cluster as background services to hide the complexities of the runtime environment from the user and interfacing to the ML software on the cluster in such a way that it will allow plugging in different ML runtime environments - beside Spark - in the future.

This article is a major extension of [34], with more details on the conceptual microservice-based architecture, related work and fundamental terms as well as technologies. It also provides additional results of a performance evaluation of the framework which are not presented in [34]. I.e., the effect of caching RDDs in Apache Spark is investigated by comparing the execution time of training and testing our benchmark evaluation models, namely Multiple Linear Regression (MLR), Decision Tree (DT), Random Forest (RF) and Gradient Boosted Trees (GBTs) models in case of caching and without caching input time series datasets. Moreover, the execution time and framework overhead are measured for evaluating the efficiency of the framework, highlighting the advantage of storing and retrieving ML models. It is also evaluated at what dataset sizes the calculation of ML models on a computing cluster outperforms calculations on single machines. To this end, the points are defined, referred as thresholds, at which a distributed computing framework based on, e.g., Apache Spark becomes necessary. This is done by comparing the total time required for training and testing different data-driven forecasting models on a computing cluster (using Apache Spark) to the time needed on a single computer for performing the same task.

The remainder of this article is organized as follows. In the next section, state-of-the-art frameworks related to our framework are presented. In Sect. 3, the fundamental terms and technologies used in the presented work are explained. In Sect. 4, the architecture of the proposed microservice-based framework is introduced. Section 5 presents the experimental evaluation of the framework and discusses the obtained results. The last section draws some conclusions and outlines future work.

2   Related Work

ML offers a variety of powerful algorithms and approaches for modeling and decision making from data, but implementing a ML model by yourself is a complex, long lasting and error prone process [12]. To ease the usage of ML, the ML community has developed a variety of powerful frameworks and tools to make its techniques more accessible to end users. Such frameworks and tools can be categorized into data analytic and ML workflow management frameworks.

Frameworks like Apache Spark which is a data analytic framework containing a good library for more traditional ML algorithms, or TensorFlow dedicated to Deep Learning, are low level frameworks that help data scientists in programming ML algorithms which could then be executed on a local computer or even for better performance on a computing cluster. Such frameworks typically don't provide easy-to-use user interfaces for non-experts by themselves but there are additional (Open Source) tools (e.g. Jupiter Notebook) which provide lean web user interfaces to such frameworks for hiding the details of the background cluster runtime environment from the user. Typically, these interfaces are aimed towards a more experienced data scientist programmer and less towards non-expert users who just want to apply ML algorithms. Apache PredictionIO [20] is an open source ML framework for developers. Besides supporting the deployment of ML algorithms, Apache PredictionIO allows expert users to train and test ML models and query results via RESTful APIs. It is built on top of state-of-the-art scalable open source services, e.g. Hadoop, HBase, Elasticsearch and Apache Spark. The drawback here is the non-existence of UI to facilitate performing ML tasks for non-expert users.

Contrary to the data analytics tools aimed for the experienced ML programmers, there are nice User Interface (UI)-based tools targeted to non-experts. Johanson et al. in [13] developed OceanTEA, a framework to analyze time series datasets in climate context. OceanTEA leverages web technology such as microservices and a nice web UI to interactively visualize and analyze time series datasets. It is a cloud-based software platform, consisting of a microservice back-end and a web UI, similar to the framework implemented in this article. Both components communicate with each other through an API gateway utilizing REST and each microservice is deployed independently through a Docker. OceanTEA provides four main UI interfaces for the exploration and analysis of oceanographic times series data including functionalities of time series data management, data exploration, spatial analysis and temporal pattern discovery.

Another project focused on the acceleration of research in energy data analysis is WattDepot presented by Brewer and Johnson in [14]. The software platform is an open source and internet-based one. It supports the collection, storage, analysis and the visualization of data coming from energy meters. The architecture encompasses three types of services, namely sensors, servers and clients. The sensors collect the data from different energy meters and send it to the services which store the incoming data by utilizing the provided RESTful APIs. Since the services are not coupled to a specific database, flexible data storage is provided. For analysis and visualization, the clients request the data from the services in the format XML, JSON or CSV. The applications of WattDepot include a web application for a dorm energy competition and a power grid simulation mechanism.

However, both WattDepot and OceanTEA typically are not generic. They contain dedicated ML based analysis features which are specialized towards the special application domain and therefore e.g. performing ML tasks such as forecasting as needed in the energy application field are not included in them.

Shrestha et al. in [15] developed a user friendly web application to analyze health and education datasets. This tool also includes ML algorithms for the forecasting of time series data. The application also has a nice and easy-to-use user interface that was developed using human-computer interaction design guidelines and principles and targeted at novice and intermediate users. The technologies used were Java, the Play framework and Bootstrap. But only linear regression, logistic regression and back propagation were utilized to perform forecasting on the input datasets. However, this framework is not able to solve ML tasks in the context of Big Data and can only be used as standalone application on a desktop computer.

ML workflow management is a rich area of research that has produced systems to manage the process of building ML models. The process of building a satisfactory ML model by a data scientist is characterized as an iterative trial-and-error procedure, where in each iteration the user reveals essential insights into the effectiveness of algorithms' configurations. Since the models may become numerous, it is important to keep track of the relevant information so a model's performance can easily be analyzed. This leads to the problem of model management which encompasses the storage and retrieval of the models and related metadata (e.g. hyperparameters, evaluation performance, etc.) in order to analyse them collectively [12].

Multiple recent research projects have been introduced addressing the model management as a part of the ML workflow. Vartak et al. in [12] introduced ModelDB, a system for tracking and versioning ML models in form of pipelines. The authors argued that data scientists are reluctant in using other environments than their favored ones, especially those with a GUI and therefore they provide native client libraries for scikit-learn and Spark MLlib which can be used to track and store models, operators and related metadata. The framework consists of a front-end and a back-end encompassing a relational database and custom storage engine. The front-end is implemented as a web UI and supports the review, inspection and comparison of the tracked and indexed models and pipelines through a Tableau-based interface. In addition, the information can be explored and analysed through SQL. The limitations here are that ModelDB is developed as a monolithic application making it difficult to be maintained and further developed. Moreover, ModelDB did not provide the ability to handle problems in the context of Big Data.

To manage ML models and their lifecycle, MLflow is introduced in [19]. Expert users can develop and track ML experiments, share and deploy ML models. MLflow is developed as an open source software system addressing typical problems of the ML workflow particularly experimentation, reproducibility and deployment. It is integrated with Python, Java and R, and provides REST APIs encompassing three main elements. The first one, MLflow Tracking, offers APIs for logging experiments and supports querying the results through APIs as well as visualizing them with a web UI. The second component, MLflow Projects, can be used to create reusable software environments for reproducibility and is defined through YAML files. The last item, MLflow Models, provides the functionality to package ML models in a generic format and deploy them. Those models incorporate similarly to MLflow Projects a YAML file which contains the metadata of the model.

To address the issue of model deployment, a variety of frameworks and tools are developed. Tensorflow serving [21] provides a flexible and powerful system for serving tensorflow models. It allows expert users to achieve an efficient integration of tensorflow models in the production environments. Kubeflow [22] is a cloud platform for ML built on top of Google's internal ML pipelines. It provides expert users with a lot of functionalities including notebooks for training and serving tensorflow models. H2O Flow [23] is another efficient framework for creating and managing ML and Deep Learning workflows including training and testing models. This framework supports Python, R and scala on top of Hadoop/Yarn and Apache Spark.

Table 2 introduces a brief comparison between the aforementioned ML frameworks based on some criteria to precisely highlight the originality of the solution proposed in the present article. In this table, data analytic frameworks are refereed as 1 and ML workflow management frameworks are referred as 2. (2.A) refers to the first sub-category of non-expert users presented in Table 1.

Table 2. Data analytic and ML workflow management frameworks.

Framework	Framework category	Web UI	Microservice architecture	Support Big Data	Support non-expert	Generic
Apache Spark	1	–	–	+	–	+
Tensorflow	1	–	–	+	–	+
Apache PredictionIO	1	–	+	+	–	+
Jupiter Notebook	1	–	+	+	–	+
OceanTea	1	+	+	–	+(2.A)	–
WattDepot	1	+	+	–	+(2.A)	–
ModelDB	2	+	–	–	+(2.A)	+
MLflow	2	+	–	–	–	+
Tensorflow serving	2	+	+	–	–	+
Kubeflow	2	–	+	+	–	+
H2Oflow	2	–	+	+	–	+
Current framework	1 + 2	+	+	+	+(2.A)	+

The framework implemented in this article uses microservice and Apache Spark, including MLlib, in addition to HDFS to provide scalability and simplicity. What differentiates the framework from the aforementioned projects, is the additional abstraction provided by the UI to support non-expert users (category A) in applying ML. Moreover, most of the aforementioned frameworks are intended and developed to mainly support expert users and do not provide an easy to use integrated framework for non-expert users. But the above tools or comparable other tools could be used as building blocks to form a more complete integrated environment such as AutoML[2] which can be seen as a competing

---

[2] https://www.automl.org/.

approach for automating the process of applying ML to real world problems. The aim of the framework presented in this article can be seen as a first step in the direction of such a more complete environment for even non-experts, designated in a way that will allow plugging in several Machine Learning and Deep Learning runtime environments.

3   Related Fundamental Terms and Technologies

In this section, we introduce the background knowledge necessary to understand the main contributions of this article.

3.1   Machine Learning

With the following definition, Alpaydin et al. in [1] introduced an essential description of machine learning: "Optimizing a performance criterion using example data and past experience". Machine learning, as its name implies, means the ability to make the computers capable to learn from data and use the resulting knowledge to perform further tasks without any guidance from the human side. Precisely, machine learning is a scientific discipline aiming at designing and developing specific algorithms and concepts in order to allow computers to evolve behaviors and react to different actions based on empirical data such as sensor data. Indeed, it can be seen as a core in the field of artificial intelligence, in which the computers can learn from existing data to predict the future behavior, results and trends.

Applying ML to extract useful knowledge from raw data has become increasingly popular in a variety of areas. One such field is the health sector where it helps with medical diagnosis [33, 42]. Virtual voice assistance, like siri and alexa, is another example, where ML is used to take voice commands from people like setting the alarm clock or finding specific information on the internet. To ensure better sustainability and economic operation of electricity grids through intelligent decision making in unit commitment of decentralized energy resources and flexible loads at grid level, an accurate prediction of future energy demand and renewable energy generation is required. To this end, ML also takes the advantage for energy load and generation forecasting [30, 31, 36–40].

Machine Learning Scenarios. Four different scenarios can be distinguished in the field of machine learning, namely supervised, unsupervised, semi-supervised and reinforcement machine learning scenarios. The main distinction between the mentioned scenarios depends on the information they handle. As a result, the behavior of learning algorithms will differ accordingly.

Supervised Machine Learning. It specifies the scenario, in which the examples in the training set are labeled with a significant information called labels. Such labels are missed in the examples in the testing set and need to be predicted [45]. More abstractly, all examples in the training set are labeled explicitly. Each of them consists of attributes or predictors on one side and the corresponding output on the other side. Both predictors and their corresponding outputs could be nominal or numeric depending on the source of data. In the supervised learning settings, we can think of a teacher who provides an extra information i.e. labels to the examples in the training set to predict such information for the unlabeled examples in the testing set.

We distinguish two different problems in the supervised machine learning scenario, namely classification and regression problems. In the current article, regression problems for time series forecasting are implemented and evaluated. Some important examples concerning the supervised machine learning scenario are:

- Linear regression for regression problems.
- Decision Trees (DT), Random Forest (RF) and Gradient Boosted Trees (GBTs) for classification and regression problems.
- Support vector machines for classification and regression problems.

Unsupervised Machine Learning. In contrast to the supervised machine learning scenario, in which the examples are explicitly labeled, the examples here are unlabeled. There is no information in the training set except the training examples containing only the features without the corresponding output [46]. The unsupervised machine learning scenario tries to discover the similar characteristics between the examples and group them into meaningful clusters. Precisely, it aims at discovering and presenting a significant structure in data. Some important examples concerning unsupervised machine learning scenario are:

- k-means algorithm as a clustering algorithm.
- Apriori algorithm as an association rule learning algorithm. This algorithm can be seen as the base of recommendation systems which try to discover the behavior of customers and present the appropriate product to them consequently.

Semi-supervised Machine Learning. It can be seen as a middle point between supervised and unsupervised machine learning scenarios [47]. In this scenario, a part of data is labeled with some supervision information i.e. labels. However, semi-supervised machine learning scenario is cheaper than the supervised one based on the fact that the labeled data is more expensive than unlabeled one. It is hard to get a labeled data because the human annotation of data is expensive and needs the utilization of experts in order to label this data. Hence, semi-supervised machine learning has gained a great advantage in different application fields.

Based on the aforementioned definition of semi-supervised machine learning scenario, the usage of unlabeled data needs some assumptions on the underling distribution of data. The main assumptions of semi-supervised machine learning scenario are:

- Smoothness assumption: in this assumption, the points that are close to each other belong to the same class.
- Cluster assumption: in this assumption, the points are clustered based on the similar characteristics between them. As a result, the points that are in the same cluster belong to the same class.
- Manifold assumption: it is commonly used with high dimensional training data, in which manifolds are learned based on labeled and unlabeled data to get rid of curse of dimensionality and then the learning process is done using distance and density within each manifold.

Reinforcement Machine Learning. In this learning scenario, the model is built based on the interaction with the environment [48]. Reinforcement machine learning scenario aims at maximizing the rewards. It differs from the supervised machine learning scenario in that the input/output pairs are not presented explicitly. On-line performance evaluation is involved in the learning process. As a result, the model will react to the evaluation feedbacks aiming at increasing the rewards and achieving the best performance. Reinforcement machine learning has become more important in the recent years, as it produces the best solutions in a lot world wide applications, for instance helicopter flying, resource-constrained scheduling, robot control systems and playing backgammon.

3.2 Big Data Technologies

With the increasing amount of available data, various libraries and systems have been introduced to enable large-scale distributed/parallel processing. One of the best known open-source frameworks is Apache Hadoop4 which supports Big Data processing and storage in a distributed environment. It encompasses various components including a distributed file system, the data processing tool MapReduce and a cluster resource manager. The Hadoop Distributed File System (HDFS) enables the reliable storage of extensive files in a cluster [25]. It provides fault tolerance by splitting the files into blocks and replicating these blocks multiple times over the cluster.

Figure 2 demonstrates the architecture of HDFS which consists of a Name Node which coordinates file system operations (e.g. opening and closing files, etc.) and multiple Data Nodes which store the file blocks and serve the read and write requests [24]. Hadoop MapReduce [27] is a programming model allowing developers to write programs to process data in parallel. Its motivation is based on the complexity related to computation parallelization, data distribution and fault tolerance. The main functions of MapReduce are Map, responsible for transforming data into key/value pairs and Reduce, which accepts the output from the Map task as input and merges matching pairs.

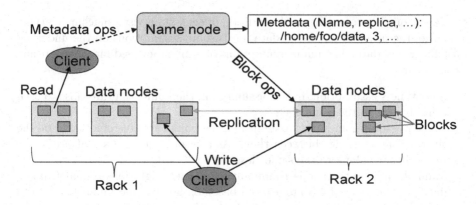

Fig. 2. HDFS architecture [24].

Yet Another Resource Negotiator (YARN) [26] is a technology that decouples the application and the required computational resources (e.g. CPUs, RAM, etc.) for processing from the resource management infrastructure of the cluster. Figure 3 illustrates YARN's architecture which is mainly composed of a Resource Manager (RM), multiple Node Managers (NM) and an Application Master (AM) for each program. When an application is submitted to the RM, the RM allocates a container accommodating the required resources for the application and contacts the related NM to launch this container. The container then executes the Application Master (AM) which coordinates the application scheduling and task execution and sends resource requests to the RM.

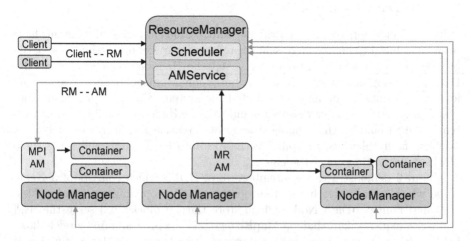

Fig. 3. YARN's architecture [26].

3.3 Microservices

Until recently, the monolithic architecture was a classic approach to implement web applications where the database, the server and the client are maintained in a single codebase. However, with the rising number of application deployments to the cloud, more and more companies like Amazon, Netflix and Zalando have shifted from a monolithic architecture to a newer more scalable architecture called Microservices. For simplicity's sake, the term service in this article refers to microservice.

Characteristics. As the name suggests, the microservices architectural style revolves around implementing an application consisting of multiple small services or entities. These services are built around the application's business functionalities, follow the Single Responsibility Principle (SRP), run in their own process and are independently deployable [32]. By following the SRP which is similar to the UNIX philosophy emphasizing programs to do one thing and doing it well, services become highly cohesive and decoupled, leading to good code maintainability. This is unlike monolithic applications which lack hard boundaries and tend to become, with added functionality, complex and tightly coupled which, in effect, leads to difficulties when changes are made since they often span multiple components.

Another distinction is that the microservices style does not require the redeployment of the whole application when new features are implemented or bugs are fixed. Instead, only the corresponding and affected service needs to be adapted and redeployed. Furthermore, microservices of a single application are not constrained to be implemented with the same set of technologies and frameworks. This allows teams working on different microservices to use independent technology stacks, as well as data storage technologies, suitable to the data they process.

Communication Types. In a microservices architecture, services are isolated from each other and distributed over a network, making communication more complicated than in monolithic applications. It is often said that microservices should have smart endpoints and dumb pipes, meaning that the logic should be inside of the services and only lightweight mechanisms and standards should be used for their communication [28]. Communication styles are usually divided into request/response and event-based techniques [29]:

- Request/response: this method describes how two services can directly communicate with each other, where one service initiates a request to another and in return expects a response.
- Event-based: this type of communication is driven by events, where one service or producer emits an event and all services that have subscribed to the event type will get an update.

REST. A common way to implement the request/response communication style is by using REST (REpresentational State Transfer), a protocol-agnostic architectural style that commonly uses HTTP as a communication protocol. All microservices implemented in this article use REST protocol to communicate between each other. This protocol enforce each service to define some RESTful APIs for transferring the data. The term REST was first coined by Fielding et al. in [2] and is made up of the following 6 constraints.

1. Client-server: to improve the portability of the client i.e. user interface and scalability of the server entities, the client and server should be separated. This constraint enables the independent involvement of both.
2. Stateless: this constraint affects the communication between the client and server and declares that it should be stateless, meaning that the client requests to the server must contain all necessary information.
3. Cache: improving the network efficiency by requiring data within a response to be labeled as cacheable or non-cacheable.
4. Uniform interface: this constraint emphasizes the importance of a uniform interface between components. To this end, the implementations are decoupled from the services they provide and the information is transferred in a standardized form rather than one which is specific to an application's needs.
5. Layered system: to simplify the complexity of an overall system, hierarchical layers should be implemented which constrict the components' behavior.
6. Code-on-demand: this is an optional constraint that allows client functionality to be extended by downloading and executing code in form of applets or scripts.

4   Concept and Architecture

In this section, the basic concepts and architecture of the proposed framework are presented. First, the general framework architecture is introduced. Then, details of the different architectural layers are presented.

4.1   Framework Architecture

Figure 4 describes the conceptual architecture of the presented framework. As seen in this figure, the architecture is layered into three main layers, namely UI layer, service layer and persistence and processing layer. The UI is split into separate sub-parts (e.g. separate web applications) providing dedicated functionalities for data and model management, model training and cluster management which are wrapped into one logical web application forming the UI of the application. The service layer is partitioned into two microservices, where each one is a small and self contained application that can be deployed independently e.g. on the runtime cluster with a single responsibility. One service focuses on data and model management, where models can be seen as special data objects.

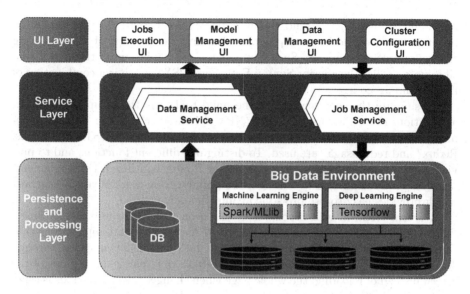

Fig. 4. Basic architecture of the proposed microservice-based framework.

The other service focuses on the management of running ML jobs e.g. for training and testing. The services provide RESTful APIs which are used by the web applications in the UI layer to interact with the runtime environment.

The persistence and processing layer provides the basic model and data storage capabilities according to the underlying runtime computer infrastructure and provides generic interfaces for executing and managing ML jobs on this infrastructure independent of the used low level ML framework. While the current implementation only supports Apache Spark as ML framework, the persistence and processing layer is designed in a way that supports plugging in additional ML frameworks in the future. In the following, the layers will be described in more details.

User Interface (UI) Layer. This layer consists of separate web applications providing dedicated functionalities which interact with the service layer via RESTful APIs. The separate web applications are wrapped into a container application which provides navigation between the views to form the complete UI. To make the user experience of the UI as pleasant as possible, the famous 10 Usability Heuristics for UI Design by Nielsen [3] are applied while conceptualizing and implementing the UI. Multiple technologies including HTML5, CSS and React[3] are utilized to implement the UI. The JavaScript (JS) library from Facebook, React, is chosen because it simplifies the development of complex user interfaces and is very permanent. Its good performance can be attributed to its use of a virtual Document Object Model (DOM) which is a copy of the HTML

---

[3] https://reactjs.org/.

DOM and enables efficient rendering updates of the otherwise slow HTML DOM. React is based on declarative programming and the concept of encapsulating and reusing of components. Such components are implemented through a specific syntax called JavaScript Syntax Extension (JSX) which is a combination of HTML and JS code.

To simplify the configuration of the build tools and the setup of the React application, Node Package Manager (NPM)[4] is used. For better data management and to organize the side effects related to asynchronous RESTful API calls, Redux[5] and redux-saga[6] are used. To distinguish different functions and to provide good navigability on the website, React Router is utilized. For implementing a responsive and nice web design, the popular framework React Bootstrap[7] which provides easy to use pre-styled components is utilized.

A recent trend in web development has been to develop web UIs as Single Page Applications (SPAs) [49]. Essentially, SPAs are front-end applications that consist of single HTML document that can be dynamically updated through JavaScript (JS). This makes it possible to refresh only particular regions of the screen instead of reloading the whole page when changes take place. This is especially convenient in interactive web pages, since these applications can respond much faster to user input and therefore provide better user experience. Additionally, the number of requests between the SPAs and services is often dramatically decreased, since much of the logic can be implemented in the front-end. For these reasons, the web UI will be implemented as an SPA communicating with the service tier through HTTP requests using the RESTful APIs. In the current version of the concept, the UI contains separate web applications for "data management", "model management", "execution of jobs" (e.g. for training and testing) and "cluster management". Figures 5 and 6 show some web page views related to these applications.

Data Management UI. It allows the uploading, management and configuration of data sources which provide data to ML jobs. Moreover, an interactive visualization besides statistical analysis can be performed on the datasets to achieve a better understanding of their characteristics and properties. For example and in the case of time series datasets, the user has the ability to zoom in/out and select a part of the chart for more detailed view. This allows the user to discover trends and outliers in the selected part of the time series dataset. Additionally, when the user hovers over a specific point in the chart, the related information will appear in a small box, for example the value of the power generation at this point. The interactive visualization of statistic and performance data in our framework is implemented using the HighChart Java-script library [50]. Moreover, dedicated features could be selected in the chosen dataset before performing ML tasks.

---

[4] https://www.npmjs.com/.

[5] https://redux.js.org/.

[6] https://redux-saga.js.org/.

[7] https://react-bootstrap.github.io/.

Model Management UI. Analog to data management, the model management UI allows the management of ML models which are (eventually) already pre-trained in the framework. Figure 6a shows a view of this UI which lists the available models. Each model has some associated metadata (e.g. id, creation date, model name, a textual description of what the mode does, etc.) which are shown in the tabular view. Each row (e.g. a pre-trained model) represents a ML pipeline corresponding to a specific ML task. For each task, the related general information resulting from performing this task such as ML algorithm, dataset used for training and testing, hyperparameters and performance results, to name a few, are shown if the user hovers over the model entry in the model list. The user can compare models and select the best one for executing it on a new dataset. Moreover, the user can perform actions on a selected model, namely delete a pipeline, extract the best hyperparameters, extract cluster configurations or extract the whole parameters and use them to build a new ML model.

Job Execution UI. It provides functionalities for executing a job for training and testing a ML model. To ease the usage for non-experts (non-programmers), the UI provides a wizard interface which guides the user through the process of choosing a dataset, a type of analysis to be performed on the dataset, an adequate ML model (e.g. model, either pre-trained or untrained) for performing the wanted type of analysis and afterwards for tuning the execution parameters of the model based on an already existing parameter set.

One of the main advantages of the proposed framework is to be very generic. I.e. in the step of selecting a given type of analysis to be performed on a dataset, the user should be able to select many different types of ML based analysis. But what kind of ML analysis methods and algorithms will be available is directly dependent on what kind of low level ML frameworks will be integrated on the persistence and processing layer.

Because in the present work only Apache Spark is integrated as low level ML framework and Apache Sparks standard ML library mainly provides algorithms for classification, clustering and regression, our framework currently only provides these three categories for choosing an analysis category as shown in Fig. 5a. After choosing one of these categories, the user will be navigated to the datasets tab view in order to select an already uploaded dataset or data source, or directly upload one to perform the ML task. Thereafter, the wizard navigates to the next wizard screen shown in Fig. 5b. Figure 5b shows that a ML framework can provide a variety of ML algorithms for performing a certain analysis category to cover a wide range of ML application scenarios. I.e., it can be seen in Fig. 5b that Apache Spark provides several algorithms for "regression analysis", e.g. "Linear regression", "Decision tree regression" and so on. If at a later time more than one ML framework will be incorporated into the present framework, different algorithms implementing an another analysis category can even be provided.

It can also be seen from Fig. 5b, that the user has the possibility to use an already existing pre-trained model or alternatively create and train a new ML model. Additionally, the user can adapt a given collection of algorithm hyper-

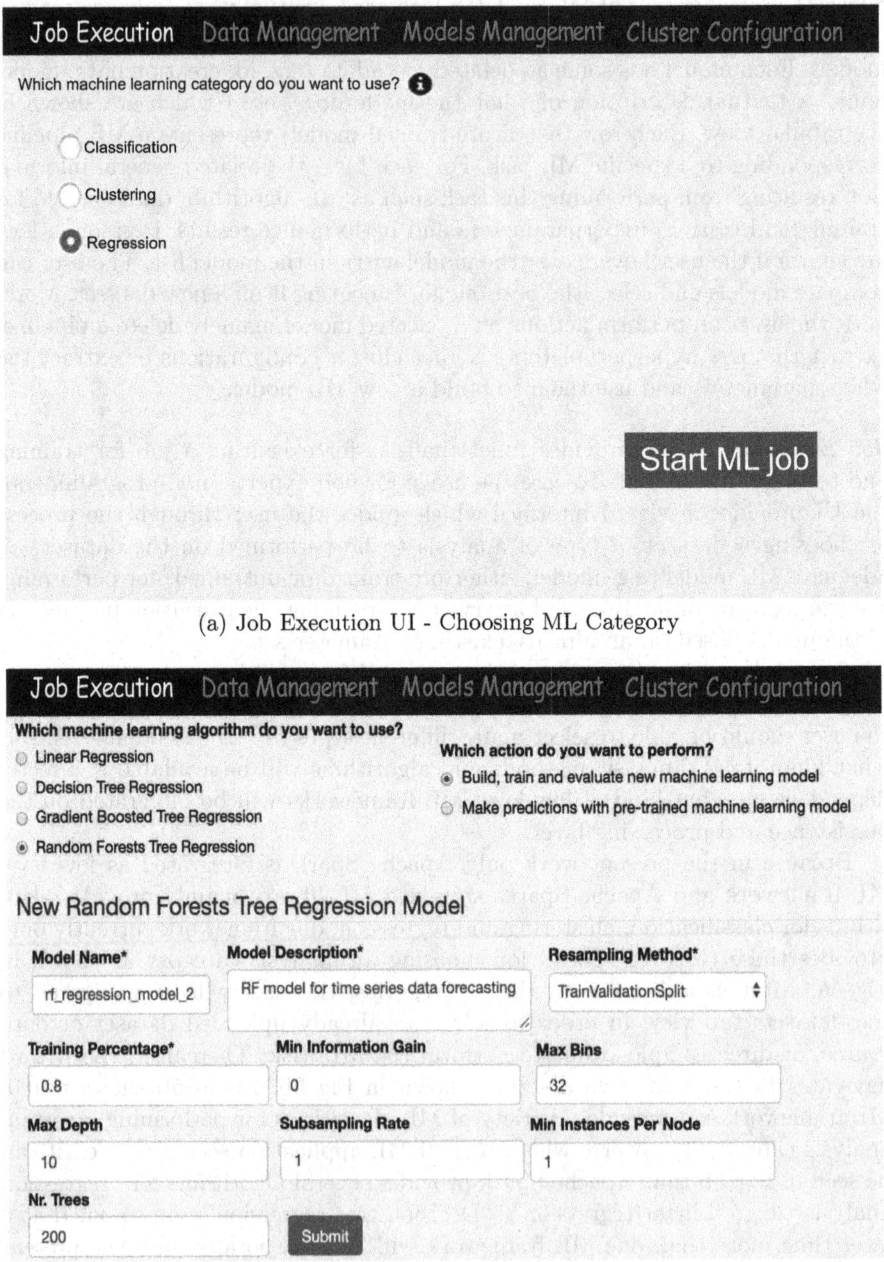

(a) Job Execution UI - Choosing ML Category

(b) Job Execution UI - Building ML Model

Fig. 5. User Interface (UI)

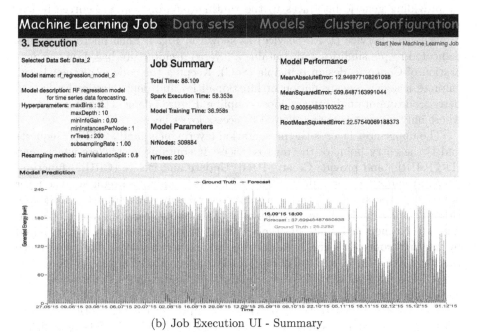

#	Created	Name	Description	Test Performance		Summary	Actions
1	16.04.2019 08:52	linear_regression_model_1	LR model for time series data forecasting.	MeanAbsoluteError: MeanSquaredError: R2: RootMeanSquaredError:	43.48 3078.24 0.40 55.48	Q	🗑 Extract HP Extract CC Extract All
2	16.04.2019 08:53	decision_tree_regression_model_1	DT model for time series data forecasting.	MeanAbsoluteError: MeanSquaredErr R2: RootMeanSquar	11.45	Q	🗑
3	16.04.2019 09:12	random_forest_tree_regression_model_1	RF model for time series data forecasting.	MeanAbsoluteEr MeanSquaredEr R2: RootMeanSquar			
4	16.04.2019 09:13	gradient_boosted_tree_regression_model_1	GBT model for time series data forecasting.	MeanAbsoluteEr MeanSquaredEr R2: RootMeanSquaredError:	22.67		Extract All

decision_tree_regression_model_1

Algorithm: Decision Tree Regression
Dataset: Data_2
Hyperparameters: maxBins : 32
maxDepth : 7
minInfoGain : 0.00
minInstancesPerNode : 1
Resampling Method: TrainValidationSplit : 0.8
Training Duration: 3.182s

(a) Model Management UI

Machine Learning Job Data sets Models Cluster Configuration

3. Execution   Start New Machine Learning Job

Selected Data Set: Data_2

Model name: rf_regression_model_2

Model description: RF regression model for time series data forecasting.
Hyperparameters: maxBins : 32
maxDepth : 10
minInfoGain : 0.00
minInstancesPerNode : 1
nrTrees : 200
subsamplingRate : 1.00

Resampling method: TrainValidationSplit : 0.8

Job Summary

Total Time: 88.109

Spark Execution Time: 58.353s

Model Training Time: 36.958s

Model Parameters

NrNodes: 309884

NrTrees: 200

Model Performance

MeanAbsoluteError: 12.946977108261098

MeanSquaredError: 509.6487163991044

R2: 0.900564853103522

RootMeanSquaredError: 22.57540069188373

Model Prediction

⊸ Ground Truth ⊸ Forecast

16.09'15 18:00
Forecast : 37.69945487650838
Ground Truth : 26.2282

(b) Job Execution UI - Summary

Fig. 6. User Interface (UI)

parameters for tuning the model performance. The storage and re-usability of pre-trained ML models on new datasets is another advantage of the presented framework. This eases usage and reduces the time the user needs to train and build a new model for each new dataset. After appropriate options are chosen in Fig. 5b, the ML task including learning and testing can be executed on the runtime platform. The wizard will then show a screen which allows to monitor the execution state. When the execution is done, the model and the other results of execution will be saved in the persistence and processing layer and a comprehensive visualization of results as well as an execution summary will be be shown as depicted in Fig. 6b.

Cluster Configuration UI. As mentioned in the introduction, a Big Data infrastructure as runtime environment for ML tasks can introduce great challenges for configuring and running the framework on the cluster with best performance for a given task. To tackle this challenge, the cluster configuration UI implemented in this framework gives the possibility to tune the low level execution framework configurations in relation to the usage of CPU cores, RAM usage and executors instances, to name a few.

Service Layer. This layer abstracts the interface of the UI applications to the ML runtime environment (e.g. computing cluster or single computer, etc.) by providing generic interfaces to the runtime environment via currently two microservices, namely the Job Management Service (J.M.-Service) and the Data Management Service (D.M.-Service) as shown in Fig. 4. Each microservice has dedicated responsibilities and contains a layered architecture based on the Separation of Concerns design principle (SoC). Keeping the code in distinct layers enforces a logical encapsulation of functionalities and dependencies leading to better code maintainability and loose coupling. Figure 7 depicts this architecture, where only upper layers are allowed to access lower layers.

The uppermost layer is the presentation layer which handles HTTP requests and is the entry point of the microservices. It contains controllers which map HTTP URLs and provide Create, Read, Update and Delete (CRUD) functionality to the outside through RESTful APIs. For simple read requests, the layer accesses the persistence layer to acquire the relevant data from the database. However, for complex logic, it communicates with the service layer which contains the business logic. This has the advantage that common operations required by multiple controllers can be abstracted to the service layer. The persistence (i.e. data access) layer consists of repositories and entities. The repositories interact with the underlying data source i.e. database and manage the entities which encapsulate the domain objects.

The following two sections provide a comprehensive description of both microservices, which are called as services for a simplicity's sake. The established RESTful pattern is chosen as the communication tool instead of the event-driven pattern, because the microservices are just two in total and the RESTful communication is easier to implement. In addition, the JSON format is selected for

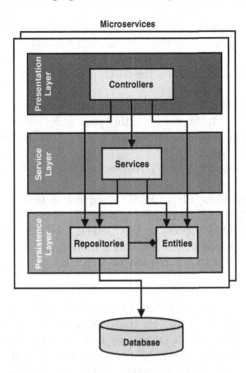

Fig. 7. Layered architecture of microservices.

requesting and sending data via the RESTful APIs because of its popularity, ease of use and interpretability.

J.M.-Service. This service is responsible for the creation and submission of jobs to be executed by an available low level ML execution framework (e.g. Apache Spark) on the available runtime environment (e.g. a cluster or single computer). Therefore, it interfaces with the persistence and processing layer below which encapsulates the specification of a certain runtime environment.

The J.M.-Service not only allows to execute ML tasks but also tracks and monitors the status of the running tasks. Moreover, it reads the execution results stored by the executing framework somewhere in the runtime environment (e.g. in an execution directory of the task on e.g. a file system) and sends them to the D.M.-Service for storage in a database, so that the execution statistics and results can be later visualized in the UI. The J.M.-Service provides an abstract job execution and monitoring interface to the web application UI through its RESTful APIs. This completely decouples the UI from the specification of the runtime environment. The main functionalities of J.M.-Service REST-APIs are described by the following URL patters:

1. **/jobs:** a GET request on this URL is used a list of spark jobs.
2. **/jobs:** a POST request on this URL is used to create of a spark job and its corresponding processing directory in HDFS.
3. **/jobs/id:** a GET request on this URL is used to retrieve a spark job for a specific id.
4. **/jobs/id:** a DELETE request on this URL is used to delete a spark job for a specific id with its corresponding processing directory in HDFS.
5. **/jobConfigurations:** a GET request on this URL is used to show a list of spark configurations.
6. **/jobConfigurations:** a POST request on this URL is used to create spark configuration.
7. **/jobConfigurations/id:** a DELETE request on this URL is used to delete a specific spark configurations for specific id.
8. **/jobSetup:** a POST request on this URL is used to copy the packaged jars and pre-trained saved machine learning models into HDFS.
9. **/submitJob/id:** a POST request on this URL is used to submit a spark job.

D.M.-Service. This service is responsible for the storage and preparation of required inputs to execute a job on the runtime environment, namely storing and providing datasets, models containing (pre-trained) algorithms and hyperparameters, to name a few. The D.M.-Service uses its own database to store the required data as well as all results produced from performing ML tasks. On the one hand, the UI applications interact with this service to upload, manage and retrieve data, model information as well as configurations. Also the J.M.-Service interacts with the D.M.-Service to retrieve information about datasets, models and configurations, copy models from the database to the execution environment of a task and to push result information back to the D.M.-Service. The D.M-Service then stores all information about the execution of a task and the results in its own database, so that these information can be later used for the visualization of the results and the overall performance of the ML jobs as already shown in Fig. 6a.

The main functionalities of the D.M.-Service REST-APIs are described in the following URL patters:

1. **/algorithms:** a GET request on this URL is used to retrieve a list of the available machine learning algorithms.
2. **/algorithms/id:** a GET request on this URL is used to retrieve a specific machine learning algorithm.
3. **/categories:** a GET request on this URL is used to retrieve a list of the available machine learning categories, for example classification, regression, clustering, to name a few.
4. **/dataSets:** a GET request on this URL is used to show available datasets
5. **/dataSets:** a POST request on this URL is used to create meta data of a dataset.
6. **/dataSets/id:** a GET request on this URL is used to retrieve the metadata of a specific dataset.

7. **/dataSets/id/data:** a POST request on this URL is used to upload a local data file into HDFS and upload the dataset's reference.
8. **/dataSets/id/descriptiveStatistics:** a POST request on this URL is used to prepare model for calculating the descriptive statistics for a specific dataset.
9. **/mlModels:** a GET request on this URL is used to retrieve a list of pre-trained machine learning models.
10. **/mlModels/id:** a GET request on this URL is used to retrieve metadata of a specific machine learning model.
11. **/mlModels/id:** a DELETE request on this URL is used to delete a specific pre-trained machine learning model.
12. **/mlModelPredictions/id:** a GET request on this URL is used to retrieve the prediction file for a specific machine learning model.
13. **/mlPipelines:** a GET request on this URL is used to retrieve a list of machine learning execution pipelines.
14. **/mlPipelines/id:** a GET request on this URL is used to get the meta data for a specific machine learning pipeline.

Persistence and Processing Layer: It hides the low level details of the runtime environment from the implementation of the services. The services use generic functions implemented in this layer to interface with the job runtime directory in HDFS and the database infrastructure installed on the runtime as well as performing dedicated tasks on the runtime environment for instrumenting installed ML frameworks to e.g. perform job execution. For each ML runtime environment, the persistence and processing layer will contain an adapter which maps model and execution details to the specific framework (see Sect. 5 for further discussion on issues related to the prototype and interfacing to the Apache Spark runtime environment).

Typically, all information related to the execution of a certain job is collected in a job runtime directory on a file system of the runtime platform. Thus, the persistence and processing layer contains functionalities for creating such directories depending on the execution framework. More generally, all data items managed by the D.M.-Service are stored in a database infrastructure which is defined by an abstract object-like interface. This interface can be implemented in the runtime infrastructure by using different database technologies as shown in Sect. 5.

5   Evaluation

So far the concept and architecture of the proposed microservice-based framework is discussed. In this section, two aspects of the experimental performance evaluation will be detailed. On the one hand, the effect of caching RDDs in Apache Spark is analyzed by comparing the execution time of training and testing the benchmark evaluation models in case of memory caching and without memory caching of the input time series datasets. On the other hand, the execution time and framework overhead for evaluating the efficiency of the framework are measured, highlighting the advantage of storing and retrieving ML models and discovering the threshold, at which the use of the proposed framework is recommended for better performing machine learning tasks in Big Data environments. Before presenting the obtained results, first the execution workflow is explained. Then the experimental setup and the related configurations are presented.

5.1   Execution Workflow

In the present work, the well-known Apache Spark framework installed on a Big Data computing cluster using an Apache Hadoop software stack as runtime engine for executing ML jobs is used. ML execution environments typically use a job runtime directory in a file system for storing all information needed for job execution (e.g. for storing models to executed, algorithm configurations and results). On a Big Data cluster based on the Apache Hadoop, HDFS is typically used as distributed file system and the runtime directory for a job can be accessed by all computing nodes of the cluster using the HDFS interfaces. Therefore, for implementing the persistence and processing layer on the cluster, HDFS and a postqreSQL database are utilized to store the required input and the output produced from performing ML tasks. The postqreSQL database system is used as an object-relational database to store all information managed by the D.M.-Service, e.g. ML categories, ML algorithms, hyperparameters, pre-trained models, jar files, references of datasets stored in HDFS, pre-trained model pipelines and untrained model pipelines.

HDFS is also utilized to store datasets and the output of successful jobs executed in Apache Spark before being read by the J.M.-Service. The dataset storage on HDFS allows it to have "Big Data" as input, i.e. datasets which are extremely large. To achieve the goal of storing pre-trained ML models in the form of binary objects, the Large Object feature of PostgreSQL is used. This feature uses the Large Object Manager Interface which stores only a reference named oid in the database table pointing to the actual object stored in the system table pg largeobject. This method breaks the binary data into chunks and allows storing objects of up to 2 GB within the database. However, another format such as Predictive Model Markup Language (PMML) will be considered in the future.

Figure 8 shows the the basic methodological workflow for task execution as it is implemented in the prototype for submitting jobs to the Apache Spark runtime. For each new job, the persistence and processing layer generates on behalf of the J.M-Service a Universally Unique Identifier (UUID) as jobID which will be sent back to the D.M.-Service. The usage of a UUID guaranties the uniqueness of the id, making it suitable to use in a distributed environment, such as a Big Data environment.

Corresponding to each jobID, a temporary job runtime directory with the UUID as a name is created in HDFS by the J.M.-Service, which uses the File

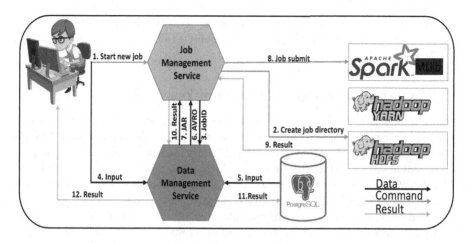

Fig. 8. Execution workflow.

System (FS) shell instruction of HDFS[8] to achieve that. Then, the J.M-Service then calls the D.M.-Service to fetch the necessary artifacts (e.g. model, runtime configuration) from the database and pass it to the J.M-Service as an Apache Spark AVRO file. After that, the J.M-Service places the AVRO file in the persistence and processing layer in the job runtime directory.

The decision for utilizing AVRO was made, because AVRO uses a schema which decouples the solution from the implementation including error prevention. An AVRO file contains the received jobID and the chosen cluster configurations. However, if no cluster configurations are chosen in the UI, the default one will be fetched from the database and used in this task. Besides cluster configurations, algorithm hyperparameters and metadata related to the execution of algorithms, namely the name of application main class are included in the AVRO file for execution. The name of the application main class is required by Apache Spark to find the main code entry point for executing the task. While all datasets are stored in the HDFS, path references pointing to the files are stored in the database of the D.M.-Service. Once the user chooses a dataset, the path reference of the dataset in HDFS is fetched from the database and included in the AVRO file. After that, the D.M.-Service fetches the corresponding jar file from the database and sends it to the J.M.-Service. At this point, all required information to perform the task is passed to the J.M.-Service which creates a spark-submit job and sends it for execution to Apache Spark.

As a result of executing e.g. a task performing forecasting on a time series dataset, the forecasting results, forecasting performance and the forecasting model in the form of a binary object are located in the temporary job runtime directory of the task. After executing the job, all of these results are stored in the temporary job runtime directory and read afterwards by the J.M.-Service to be passed to the D.M.-Service. The D.M.-Service receives the results and stores

---

[8] https://hadoop.apache.org/docs/stable/.

them in the form of a pipeline in the database to be retrieved later. Simultaneously, the D.M.-Service sends the results to the UI to be rendered and visualized for the user.

5.2 Experimental Setup and Configurations

The aforementioned microservice architecture is implemented using Java and tested while running on a local workstation which is a MacBook with a 2.7 GHz Intel Core i5 processor and 8 GB of RAM. Both microservices are implemented as standalone Spring Boot applications which are configured to run on different HTTP ports, namely 8090 and 8080. To run our web application, the embedded Apache Tomcat server from Spring Boot is utilized.

For our evaluation and to investigate the effectiveness of our framework, local execution context and cluster execution context have been configured. In the local context, Spark (v. 2.3.0) on top of Hadoop (v. 2.7.6) as state-of-the-arts technology to perform machine learning tasks is installed on the aforementioned workstation, where the executors and drivers run in a single JVM. In the cluster context, we utilize a powerful Big Data stack, in which Apache Spark is fit on top of Yet Another Resource Negotiator (YARN) as a resource manager and Hadoop Distributed File System (HDFS) as a primary data storage. The Big Data stack is deployed on a cluster of 3 logical machine nodes. Each of them has 32 cores and 80.52 GB RAM. The nodes are connected to each other by a LAN with 10 GBit/s bandwidth.

Table 3. Default and custom configurations used in cluster context.

Default	Custom
Drivers.cores = 1	Drivers.cores = 1
Driver.memory = 1 GB	Driver.memory = 1 GB
Executors.cores = 2	Executors.cores = 2
Executors.memory = 1 GB	Executors.memory = 70 GB
Executors.instances = 1	Executors.instances = 3

In the cluster context, we distinguish two configuration setups, namely default and custom as presented in Table 3. Random Forest (RF), Multiple Linear Regression(MLR), Gradient Boosted Trees (GBTs) and Decision Tree (DT) are used as base classifiers to build the data-driven forecasting models. MLlib, which is a Spark's scalable ML library is employed to build the models. To train and test the forecasting models, a simulated energy multivariate time series dataset is used. MLR is a widely used supervised algorithm which assumes a linear relationship between one or multiple independent input variables and a dependent output variable [44]. Table 4 presents the default values of the MLR hyperparameters.

Table 4. Default hyperparameters of MLR algorithm in MLlib.

Hyperparameter	Description	Default
maxIter	Maximum number of iterations	100
regParam	Regularization/Shrinkage parameter	0.0

DT algorithm [44] is a supervised algorithm, often chosen for its interpretability. It has the ability to capture the non-linear structures in data, unlike MLR. A DT is essentially a binary tree which recursively partitions the input space and consists of internal nodes and leaves (i.e. terminal nodes). It is constructed starting from the root and its nodes are split down based on the largest decrease in impurity. For classification trees, the impurity is often measured with the Gini impurity or entropy. However, for regression trees, where the target is continuous, the impurity is based on variance reduction. Table 5 presents the default values of the DT hyperparameters. RF algorithm [44] builds a forest of multiple DTs that are independently trained. Whereas, single DTs are often said to overfit, the RF algorithm does not overfit because of the Law of Large Numbers [7]. Also, randomness is applied to the training process of RF by utilizing random feature subsets for node splitting. Since, each DT is trained separately, multiple trees can be trained in parallel. For the final prediction, the individual votes of all trees are combined. Table 6 presents the default values of the RF hyperparameters.

Table 5. Default hyperparameters of DT algorithm in MLlib.

Hyperparameter	Description	Default
maxBins	Maximum number of bins for split decision and discretization of continuous features	32
maxDepth	Number of trees in the forest	5
minInstancesPerNode	Minimum number of trees (training instances) in children must have by splitting	1

Table 6. Default hyperparameters of RF algorithm in MLlib.

Hyperparameter	Description	Default
maxDepth	Maximum depth of individual trees in the forest	5
numTree	Number of trees in the forest	20

In contrast to RF which trains the trees independently, GBTs algorithm [44] employs the Boosting technique training one tree at a time. Successively, to correct the errors made by previous trees, a DT is fitted on the residuals of the previous tree, instead of a fraction of the original data. The final prediction is based on a weighted majority vote. Table 7 presents the default values of the GBTs hyperparameters.

Table 7. Default hyperparameters of GBTs algorithm in MLlib.

Hyperparameter	Description	Default
maxDepth	Maximum depth of the individual trees	5
maxIter	Maximum number of iterations	20
stepSize	Controls the contribution/weight of each tree	0.1
subsamplingRate	Training data proportion used for learning each tree	1.0

Tuning hyperparameters is an important step of the Machine Learning Pipeline (MLP), since they can not only significantly influence the forecasting performance of a model, which is not our focus in the present work, but also the processing time.

Table 8. ML algorithms hyperparameters after tuning.

ML algorithm	Hyperparameters
Multiple Linear Regression (MLR)	Max iterations (ntree) $= 20$ Regularization parameter $= 0.5$
Decision Tree (DT)	Max bin $= 5$ Max depth $= 5$ Min instance split $= 1$
Gradient Boosted Trees (GBTs)	Max depth $= 5$ Number of trees $= 20$ Step size $= 0.1$ Sampling rate $= 1.0$
Random Forest (RF)	Max depth $= 5$ Number of trees (ntree) $= 20$

Based on the main property of our microservice-based framework in facilitating training and testing ML models in Big Data environments, an efficient hyperparameter tuning is performed for the aforementioned ML algorithms to ensure that the time measurements are taken for a best case scenario of the aforementioned algorithms. The results are depicted in Table 8.

As mentioned before, one of the main advantages of the proposed framework is to store pre-trained models in order to use them later in production. Thus, for evaluation, two execution contexts are determined, namely the untrained model pipeline and pre-trained model pipeline. In the first one, as its name implies, the user follows the general methodology to perform a ML task, in which the model is trained from scratch and afterwards tested. In the second one, the user selects a pre-trained model from the database and uses it to perform or test a ML task with a new dataset without the need for building a new model. In the present article, the main goals of evaluation are discovering the effect of caching in Apache Spark, the advantage of storing ML models and reusing them, measuring the framework overhead and determining the thresholds for efficiently performing ML tasks on the cluster. To this aim, time measurements need to be precisely defined. As time measurements, we defined T_{total} and T_{fo} according to Eq. 1 and 2 respectively.

$$T_{total} = T_{exe} + T_{fo} \tag{1}$$

where:

- T_{exe}: is the execution time required by Apache Spark to perform a ML task in context of pre-trained pipelines or untrained pipelines.
- T_{fo}: is the framework overhead.

$$T_{fo} = T_{co} + T_{dbo} \tag{2}$$

where:

- T_{co}: describes the communication overhead between microservices and inside the Big Data infrastructure.
- T_{dbo}: describes the overhead for storing and retrieving required data from the database.

5.3   Experimental Results and Analysis

In the following, the evaluation results are discussed. As the focus is on the execution time and the framework overhead raised while performing ML tasks, the accuracy of forecasting will not be taken into account.

Effect of Caching in Apache Spark. Resilient Distributed Datasets (RDDs) are the basic data structure of Apache Spark developed as a fault-tolerant immutable collection of objects which can be computed on different nodes of the cluster[9]. Caching RDDs in Apache Spark is a widely used mechanism for speeding up the running applications. This is especially helpful, when running iterative machine learning applications, where the data is accessed repeatedly. If RDD is not cached, nor checkpointed, it is re-evaluated again each time an action is invoked on that RDD. The training time is measured as the time it takes to fit the model on the training data. The prediction time is similarly computed for applying the resulted model on testing data. Since Spark utilizes lazy evaluation for data transformations, meaning an operation is not executed until an action is called on the data, the prediction time has to be measured in combination with performing an action. The main advantages of the lazy evaluation mechanism in Apache Spark are:

---

[9] https://spark.apache.org/docs/latest/rdd-programming-guide.html.

– Increased manageability of RDDs because the source code of our machine
learning algorithms is organized into smaller operations which in turns
reduces the number of passes on data by grouping the operations.
– More efficient computation time and an increased speed, as only the necessary
values are computed saving the communication round-trip time between the
drivers and clusters.
– Better optimization of operations on data by reducing the number of queries.

Table 9. Mean computation time for training and testing different algorithms in the
cases of caching and no caching of input data.

Machine learning algorithms	Training time (s)		Prediction time (s)	
	No caching	Caching	No caching	Caching
Multiple Linear Regression (MLR)	16.07	3.57	3.92	0.87
Decision Tree (DT)	15.88	3.21	3.41	0.86
Gradient-boosted trees (GBTs)	37.04	21.74	8.61	1.77
Random Forest (RF)	23.11	12.48	5.75	1.12

Table 9 shows how caching of the input time series datasets affects the performance of the implemented algorithms, using their default hyperparameters
and default cluster configurations. For calculating these values, the experiments
are repeated three times. Afterwards, the mean values are calculated as final
performance indicator. Obviously, the need for caching will be larger in the case
of large datasets, as more operations are required and larger amount of data are
loaded and accessed repeatedly, therefore and to precisely discover the effect of
caching, the models are trained and tested on a small dataset size i.e. 4 MB. As
shown in this table, combining lazy evaluation with caching reduces the training
and prediction time of all algorithms by approximately 75%.

Advantage of Storing and Retrieving ML Models. The main ML task
used for this part of evaluation is to perform short-term energy generation forecasting using MLR, RF, DT and GBTs data-driven models on simulated energy
multivariate time series dataset. The algorithm hyperparameter configurations
shown in Table 8 are used. For better utilization and exploitation of the available
abilities of the underlying Big Data cluster, the custom configurations shown in
Table 3 are used. A feature space consisting of 5 features, namely temperature,
humidity, cloud coverage, hour and day is used to build the forecasting models.
A dataset of 4 GB size is used for training and testing ML models, where 80% of
the input time series dataset are used as a training set and 20% as testing set.
For each ML algorithm, the experiment is repeated three times. Afterwards, the
mean values are calculated as final performance indicator. Figure 9 shows the
T_{total} required by the framework to perform the aforementioned task in case of
pre-trained and untrained model pipeline.

In general, the total time T_{total} is strongly related to the complexity of ML models. As this complexity increases, T_{total} required to perform the task will dramatically increases. The base classifier of both RF and GBTs algorithms is the DT algorithm. Consequently, the complexity of RF and GBTs models will be higher than the complexity of the DT model. As seen in Fig. 9, RF and GBTs introduced higher T_{total} than DT and MLR algorithms.

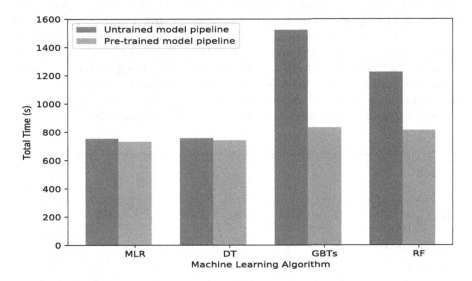

Fig. 9. T_{total} required for training and testing models (untrained model pipeline) and for testing (pre-trained model pipeline) on simulated energy multivariate time series dataset with size 4 GB.

Both GBTs and RF are algorithms for learning ensembles of trees, but the training processes are different. While GBTs algorithm trains one tree at a time, RF algorithm can train multiple trees in parallel. This can be seen clearly in Fig. 9, in which GBTs show higher T_{total} than RF. In our experiments, both MLR and DT algorithm introduce lower T_{total} compared to RF and GBTs. The efficiency of storing ML model can clearly be seen in case of complex ML models, namely GBTs and RF models, and will rise with growing complexity of the model. As the complexity of model increases, the time needed to perform the same task with each new dataset will dramatically increase and the benefit of using pre-trained models will also increase. E.g., by performing forecasting, we gain a time of 690 and 411 s in case of GBTs and RF respectively. In contrast to that, only small time will be gained in case of retrieving and reusing simpler models such as MLR and DT as seen in Fig. 9.

As a result, the recommendation of storing ML models and reusing them in testing is higher in case of complex models than for simpler ones. This experimental study gives an evidence for the importance of storing and retrieving ML models as a major property in our framework. However, the experiments are performed only with a dataset of 4 GB size. As this size increases, the complexity of the ML models will increase too, paving the road to save and gain more time for performing ML tasks with new datasets based on pre-trained models without the need for training these models.

Framework Overhead. To evaluate the framework overhead, MLR models for short-term energy generation forecasting are used. The algorithm hyperparameter configurations shown in Table 8 besides the custom cluster configurations are used in this group of experiments. The evaluation instruments the untrained model pipeline, in which the training and the testing steps of ML models are required. The goal of the study is to evaluate the effect of input dataset size on framework performance in the form of framework overhead defined in Eq. 2. For this evaluation, the size of the input datasets is upscaled to 64 GB, as bigger datasets typically expose more load on the framework infrastructure.

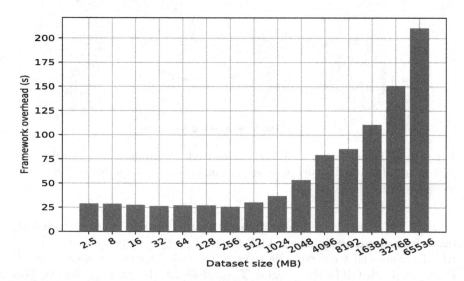

Fig. 10. Effect of input datasets size used for training and testing MLR models on the framework overhead.

As defined in Eq. 2, the framework overhead encompasses communication overhead and database overhead. The obtained results depicted in Fig. 10 show that the proposed framework introduces an approximately constant communication overhead averaging at around 26 s for datasets with sizes up to 512 MB. The framework overhead starts to increase for a size of input datasets larger than 512 MB. The reason behind this is the additional overhead raised inside the Big Data environment for resource scheduling, coordination and network communications in the cluster. Precisely, an increasing size of the input dataset naturally leads to an increased overhead due to data replication, disk I/O and the

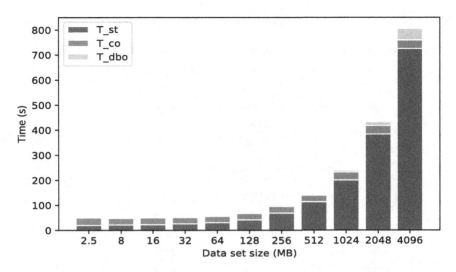

Fig. 11. Effect of input datasets size used for training and testing MLR models on the framework overhead (detailed overview).

serialization of data inside the execution environment of the cluster. A detailed increasing in overhead can be seen also in Fig. 11.

Despite this increment, the introduced framework overhead is still low compared to the execution time spent in performing a ML task as shown in Table 10. For example, the portion of framework overhead is 210,47 s in the worst case, namely for 65 GB input multivariate time series datasets. Consequently, our evaluation demonstrates, that it maintains high performance ML processing with low framework overhead to facilitate and solve ML tasks in Big Data environments, where the user gains great benefits from reusing pre-trained models.

Cluster Utilization Threshold. This section discusses the question "when to use the proposed framework for performing Ml tasks more efficiently on a cluster?". Clearly, the dataset size has an essential effect on the complexity of machine learning models and therefore on runtime performance. As the size of the dataset used for training and testing machine learning models grows, the complexity of model will increase which dramatically affects the total execution time in our microservice-based framework. While MLR forecasting models have the lower complexity, the RF forecasting models represent the higher complex models in our evaluation study. Moreover, DT forecasting model has higher and lower complexity from LR and GBT respectively as seen in Fig. 9.

The algorithm hyperparameter configurations shown in Table 8 are used. The input dataset size is changed between 2.5 MB and 4 GB in the experiments for investigating the effect of dataset size on the framework overhead and execution

Table 10. Execution time in Apache Spark vs. framework overhead for MLR models.

DataSet size (MB)	Execution time (s)	Framework overhead (s)
4	19,99	28, 78
8	21,79	28,53
16	22,6	27,41
32	25,88	26,21
64	30,54	27,05
128	41,7	27,13
256	68,54	25,82
512	113,25	30,2
1024	200,66	37,06
2048	383,75	53,54
4096	724,98	79,37
8192	1016,48	85,65
16384	4724,98	110,66
32768	6383,75	150,88
65536	11804,36	210,47

time. The total time T_{total} is compared to the time required for performing the same task in local and cluster context. The ratio of local time and cluster time is defined as *abs_threshold* in Eq. 3.

$$abs\_threshold = \frac{T_{local}}{T_{total}} \qquad (3)$$

where:

- T_{local}: encompasses the total time required to locally execute a machine learning task.

The main idea behind defining *abs_threshold* is to find the dataset size for which the total time in local context exceeds the total time required by the framework to execute tasks in cluster context. From this point, it is highly recommended to use a cluster. Precisely, to effectively perform machine learning tasks, this ratio should be greater than 1.

Performing machine learning tasks in cluster context introduces additional overhead. The main reason behind this interest lies in the time cost for resource scheduling, coordination and network communications in the cluster. Figure 12 shows the mean total time in local and cluster modes, including default and custom configurations, using various dataset sizes. It can be observed that enlarging the dataset size from 2.5 MB to 64 MB has no significant effect on both T_{local} and T_{total}.

As seen in Fig. 12a and for data less or equal to 256 MB, T_{total} in both cluster modes is larger than T_{local} in local mode which can be explained by

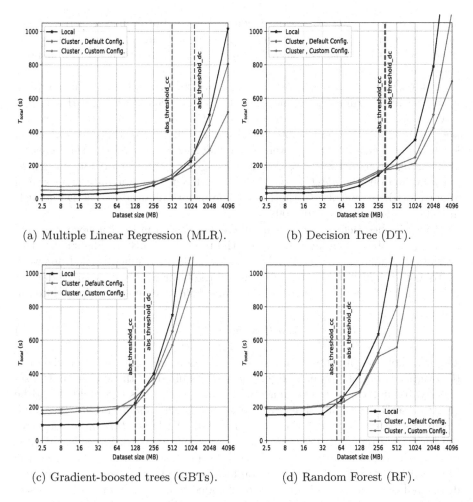

(a) Multiple Linear Regression (MLR).

(b) Decision Tree (DT).

(c) Gradient-boosted trees (GBTs).

(d) Random Forest (RF).

Fig. 12. Mean T_{total} in case of local and cluster (default, custom) configurations mode to determine the cluster utilization *abs_threshold* for MLR, DT, RF and GBTs algorithms.

the added overhead for processing the application on the cluster. Thus, running Spark applications locally for these dataset sizes is more efficient. For a dataset size less than 256 MB, T_{total} with custom configurations is larger than T_{total} with default configurations, since two additional nodes are used in these configurations where each of them introduces an overhead. As expected, when the dataset size grows larger, utilizing a cluster becomes more desirable which is shown by the intersection points highlighted by the two red lines, where these points depend on the configurations. As the *abs_threshold_cc* (cc: custom configurations) is found at a dataset size of 512 MB making the custom configurations the most efficient beyond that point, the *abs_threshold_dc* (dc: default configurations) lies at a dataset size of $>= 1024$ MB. Consequently, the computing power of the cluster can be seen and the time consumed locally to perform a task will exceed the time required to perform the same task on cluster. Therefore, it is recommended here to use the cluster.

Comparing Figs. 12a, 12b, 12c and 12d, we conducted that as the complexity of machine learning model increases, the *abs_threshold* will be early arrived. The reason is that the complex models need more calculation costs. As a result, the performance in cluster context will earlier outperform the performance in local context because of the power of the underlying deployed Big Data cluster. Concerning RF model which represents the highest complex model in our benchmark evaluation, the *abs_threshold* will be arrived for input dataset in size of about 64 MB. In contrast to that, lower complexity models such as DT models introduced *abs_threshold* for 300 MB.

As mentioned before, there is an inherent overhead in the framework arising from e.g. database communication and the use of the cluster. The smaller this overhead is compared to Spark's execution time, the more efficient the framework is. To gain insight into how the efficiency of the framework varies as the dataset grows, a new threshold, referred as *min_threshold*, is defined and formulated in Eq. 4:

$$min\_threshold = \frac{T_{exe}}{T_{fo}} \qquad (4)$$

This threshold is defined based on the fact that for an efficient execution of a task, the overhead time should not exceed the time required for the execution. Consequently, to effectively perform machine learning tasks, this threshold should be greater than 1. The main difference between *min_threshold* and *abs_threshold* lies in the context, in which they are calculated. While *abs_threshold* compares the total time required to perform a machine learning task in local and cluster context, the other one is calculated only in cluster context comparing the framework overhead with the execution for different dataset size. As a result, we discovered the point at which it is recommended to use our framework in cluster context.

This group of experiments are conducted using default cluster configurations summarized in Table 3 and also repeated three time for more robust results. The obtained mean results, presented in Fig. 13 show that *min_threshold* has the same behavior of *abs_threshold*. It is evident that for very small dataset sizes *min_threshold* is less than 1 since more time is spent on T_{fo} than T_{exe}. The *min_threshold* comes closest to 1 at the size of 64 MB and 32 MB for MLR and DT respectively as seen in Figs. 13a and 13b. Beyond this point T_{exe} starts to exceed T_{fo} which implies that for larger dataset sizes it is recommended to use the framework in cluster context. Precisely, the gap between T_{fo} and T_{exe} increases proportionally to the dataset size, since T_{exe} is strongly dependent on it. As the complexity of the model increases, the *min_threshold* will be shifted to meet smaller dataset size i.e. 2.5 MB as seen in Figs. 13c and 13d. Combining the results of *abs_threshold* and *min_threshold*, it is recommended to perform

ML tasks using the proposed microservice-based framework if both of these thresholds are greater than 1.

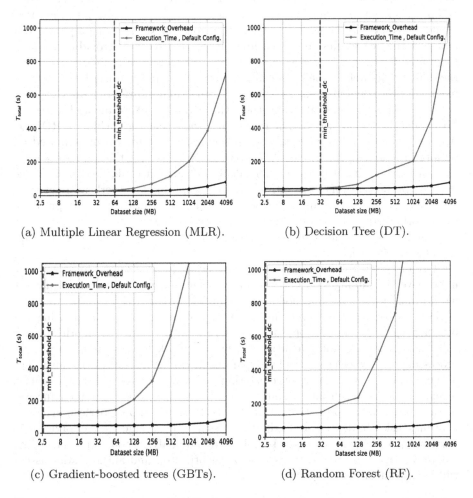

(a) Multiple Linear Regression (MLR).   (b) Decision Tree (DT).

(c) Gradient-boosted trees (GBTs).   (d) Random Forest (RF).

Fig. 13. Mean T_{total} in case of local and cluster (custom) configurations mode to determine the cluster utilization $min\_threshold$ for MLR, DT, RF and GBTs algorithms.

6   Conclusion and Future Works

The present paper introduces a new highly scalable generic microservice-based framework to ease and streamline the performing of ML tasks in Big Data environments. This framework provides a user-friendly UI built on top of a service layer that eases the usage of ML frameworks on Big Data infrastructure and hides cluster details from the user. Despite the ability of training, testing, managing, storing and retrieving of machine learning models in the context of Big Data, the framework provides functionalities for uploading, exploring and visualizing datasets using state-of-the-arts web technologies. Moreover, ML model selection and management in form of storing pipelines are supported. Each pipeline corresponds to a specific ML task, in which ML algorithm, dataset, hyperparameters and performance results are stored in an integrated package providing the user the ability for deeper comparison and better selection of ML models. To reduce the difficulty as well as the complexity of performing tasks in Big Data environments, cluster configurations can be easily tuned and adjusted in the UI.

In a comprehensive evaluation study, the advantage of storing and retrieving ML models is demonstrated. The results also show that the caching of RDDs in Apache Spark plays an essential role in saving the execution time required for performing the task on the cluster. Moreover, by measuring the framework overhead and comparing it to the model calculation time, it could be demonstrated that the proposed framework introduces an acceptable low overhead relative to an increasing size of an input dataset. For efficient utilization of the proposed framework, certain thresholds are defined to determine the dataset size, in which it is highly recommended to use clusters in favor to single computers for performing a given ML task.

The proposed framework is an ongoing work for developing an even more interactive and intelligent framework for fully automating, managing, deploying, monitoring, organizing, and documenting of ML tasks.

Future work will discover the effect of caching in the case of using the best hyperparameters that optimize the performance of the ML algorithms. We will also extend the functionalities of the framework to cover automated preprocessing, model selection and hyperparameter tuning leveraging the advantage of meta learning. Classification, clustering and a wide range of ML application scenarios will be taken into account. Deep Learning as a pluggable engine will be integrated in the persistence and storage layer to support performing Deep Learning tasks. In-depth user feedback assessment by a large number of users, in particular, non-expert users will be collected and analyzed too. For tenancy and secure management of ML tasks, user authentication and authorization issues will be also taken into account.

References

1. Vernon, V.: Implementing Domain-Driven Design, p. 612. Addision-Wesley, Upper Saddle River (2013)
2. Fielding, R.T.: Architectural Styles and the Design of Network-Based Software Architectures. AAI9980887. University of California, Irvine (2000)
3. Nielsen, J.: 10 usability heuristics for user interface design. Nielsen Norman Group **1**, 1 (1995)
4. Sebastiani, F.: Machine learning in automated texT categorization. ACM Comput. Surv. (CSUR) **34**(1), 1–47 (2002)
5. Padmanabhan, J., Johnson Premkumar, M.J.: Machine learning in automatic speech recognition: a survey. IETE Tech. Rev. **32**, 1–12 (2015)

6. Kononenko, I.: Machine learning for medical diagnosis: history, state of the art and perspective. Artif. Intell. Med. **23**(1), 89–109 (2001)
7. Voyant, C., et al.: Machine learning methods for solar radiation forecasting: a review. Renew. Energy **105**, 569–582 (2017)
8. Jurado, S., Nebot, A., Mugica, F., Avellana, N.: Hybrid methodologies for electricity load forecasting: entropy-based feature selection with machine learning and soft computing techniques. Energy **86**, 276–291 (2015)
9. Gandomi, A., Haider, M.: Beyond the hype: Big Data concepts, methods and analytics. Int. J. Inf. Manag. **35**(2), 137–144 (2015)
10. Karun, A.K., Chitharanjan, K.: A review on Hadoop-HDFS infrastructure extensions. In: 2013 IEEE Conference on Information and Communication Technologies, pp. 132–137. IEEE (2013)
11. Nadareishvili, I., Mitra, R., McLarty, M., Amundsen, M.: Microservice Architecture: Aligning Principles, Practices and Culture. O'Reilly Media Inc. (2016)
12. Vartak, M., et al.: Model DB: a system for machine learning model management. In: Proceedings of the Workshop on Human-In-the-Loop Data Analytics, p. 14. ACM (2016)
13. Johanson, A., Flogel, S., Dullo, C., Hasselbring, W.: OceanTEA: exploring ocean-derived climate data using microservices (2016)
14. Brewer, R.S., Johnson, P.M.: WattDepot: an open source software ecosystem for enterprise-scale energy data collection, storage, analysis and visualization. In: 2010 First IEEE International Conference on Smart Grid Communications. 2010 1st IEEE International Conference on Smart Grid Communications (SmartGridComm), pp. 91–95, Gaithersburg, MD, USA. IEEE (2010)
15. Shrestha, C.: A web based user interface for machine learning analysis of health and education data (2016)
16. Schelter, S., Böse, J.-H., Kirschnick, J., Klein, T., Seufert, S.: Automatically tracking metadata and provenance of machine learning experiments (2017)
17. Obe, R.O., Hsu, L.S.: PostgreSQL: Up and Running: a Practical Guide to the Advanced Open Source Database. O'Reilly Media Inc. (2017)
18. Meng, X., et al.: MLlib: machine learning in Apache Spark. J. Mach. Learn. Res. **17**(1), 1235–1241 (2016)
19. Zaharia, M., et al.: Accelerating the machine learning lifecycle with MLflow. IEEE Data Eng. Bull. **41**(4), 39–45 (2018)
20. Chan, S., Stone, T., Szeto, K.P., Chan, K.H.: Predictionio: a distributed machine learning server for practical software development. In: Proceedings of the 22nd ACM International Conference on Information and Knowledge Management, pp. 2493–2496. ACM (2013)
21. TensorFlow Serving. https://www.tensorflow.org/serving. Accessed 4 Feb 2020
22. kubeflow. https://www.kubeflow.org/. Accessed 4 Feb 2020
23. Candel, A., Parmar, V., LeDell, E., Arora, A.: Deep Learning with H2O. H2O. AI Inc. (2016)
24. Borthakur, D.: The Hadoop distributed file system: architecture and design. In: Hadoop Project Website, vol. 11, p. 21.0 (2007)
25. Shvachko, K., Kuang, H., Radia, S., Chansler, R.: The Hadoop distributed file system. In: 2010 IEEE 26th Symposium on Mass Storage Systems and Technologies (MSST). 2010 IEEE 26th Symposium on Mass Storage Systems and Technologies (MSST), Incline Village, NV, USA, pp. 1–10. IEEE, May 2010
26. Vavilapalli, V.K., et al.: Apache Hadoop YARN: yet another resource negotiator. In: Proceedings of the 4th Annual Symposium on Cloud Computing - SOCC 2013. The 4th Annual Symposium, pp. 1–16. ACM Press, Santa Clara (2013)

27. Dean, J., Ghemawat, S.: MapReduce: simplified data processing on large clusters. Commun. ACM **51**(1), 107–113 (2008)
28. Microservices. https://martinfowler.com/articles/microservices.html. Accessed 18 Feb 2020
29. Newman, S.: Building Microservices: Designing Fine-Grained Systems, 1st edn. O'Reilly Media, Beijing (2015)
30. Coughlin, K., Piette, M., Goldman, C., Kiliccote, S.: Estimating demand response load impacts: evaluation of base line load models for non-residential buildings in California. Technical report, Ernest Orlando Lawrence Berkeley National Laboratory, Berkeley, CA, USA (2008)
31. Khotanzad, A., Afkhami-Rohani, R., Lu, T.L., Abaye, A., Davis, M., Maratukulam, D.J.: ANNSTLF-a neural-network based electric load forecasting system. IEEE Trans. Neural Netw. **8**(4), 835–846 (1997)
32. Evans, E.: Domain-Driven Design: Tackling Complexity in the Heart of Software, p. 529. Addison-Wesley, Boston (2004)
33. Shoeb, A.H., Guttag, J.V.: Application of machine learning to epileptic seizure detection. In: ICML (2010)
34. Shahoud, S., Gunnarsdottir, S., Khalloof, H., Duepmeier, C., Hagenmeyer, V.: Facilitating and managing machine learning and data analysis tasks in Big Data environments using web and microservice technologies. In: Proceedings of the 11th International Conference on Management of Digital EcoSystems, pp. 80–87 (2019)
35. Witten, I.H., Frank, E., Hall, M.A., Pal, C.J.: Data Mining: Practical Machine Learning Tools and Techniques. Morgan Kaufmann (2016)
36. Aman, S., Simmhan, Y., Prasanna, V.K.: Improving energy use forecast for campus micro-grids using indirect indicators. In: 2011 IEEE 11th International Conference on Data Mining Workshops. IEEE, pp. 389–397 (2011)
37. Hong, T., Gui, M., Baran, M., Willis, H.: Modeling and forecasting hourly electric load by multiple linear regression with interactions. In: IEEE PES General Meeting. IEEE, pp. 1–8 (2010)
38. Metaxiotis, K., Kagiannas, A., Askounis, D., Psarras, J.: Artificial intelligence in short term electric load forecasting. Energy Convers. Manag. **44**(9), 1525–1534 (2003)
39. Mori, H., Takahashi, A.: Hybrid intelligent method of relevant vector machine and regression tree for probabilistic load forecasting. In: 2011 2nd IEEE PES International Conference and Exhibition on Innovative Smart Grid Technologies, pp. 1–8. IEEE (2011)
40. Cui, C., Wu, T., Hu, M., Weir, J.D., Li, X.: Short-term building energy model recommendation system: a meta-learning approach. Appl. Energy **172**(2016), 251–263 (2016)
41. Mitchell, T.M.: Machine Learning. McGraw-Hill Series in Computer Science, 414 pp. McGraw-Hill, New York (1997)
42. Cruz, J.A., Wishart, D.S.: Applications of machine learning in cancer prediction and prognosis. Cancer Inform. **2**, 59–77 (2006)
43. Breiman, L.: Random forests. Mach. Learn. **45**(1), 5–32 (2001)
44. Machine Learning Library (MLlib) Guide. https://spark.apache.org/docs/latest/ml-guide.html. Accessed 19 Feb 2020
45. Dougherty, J., Kohavi, R., Sahami, M.: Supervised and unsupervised discretization of continuous features. In: Proceedings of the Twelfth International Conference on Machine Learning, vol. 12, pp. 194–202 (1995)
46. Hahne, F., Huber, W., Gentleman, R., Falcon, S.: Bioconductor Case Studies. Springer, New York (2010). https://doi.org/10.1007/978-0-387-77240-0

47. Chapelle, O., Scholkopf, B., Zien, A.: Semi-supervised learning. IEEE Trans. Neural Netw. **20**(3), 542–542 (2009). (Chapelle, O. et al. (eds.) (2006)) (bibbook reviews)
48. Kaelbling, L., Littman, M., Moore, A.: Reinforcement learning: a survey. J. Artif. Intell. Res. **4**, 237–285 (1996)
49. Mikowski, M., Powell, J.: Single Page Web Applications: JavaScript End-to-End. Manning Publications Co. (2013)
50. Kuan, J.: Learning Highcharts. Packt Publishing Ltd. (2012)

Stable Marriage Matching
for Homogenizing Load Distribution
in Cloud Data Center

Disha Sangar, Ramesh Upreti, Hårek Haugerud, Kyrre Begnum,
and Anis Yazidi[✉]

Autonomous Systems and Networks Research Group,
Department of Computer Science, Oslo Metropolitan University, Oslo, Norway
anisy@oslomet.no

Abstract. Running a sheer virtualized data center with the help of *Virtual Machines* (VM) is the de facto-standard in modern data centers. Live migration offers immense flexibility opportunities as it endows the system administrators with tools to seamlessly move VMs across physical machines. Several studies have shown that the resource utilization within a data center is not homogeneous across the physical servers. Load imbalance situations are observed where a significant portion of servers are either in overloaded or underloaded states. Apart from leading to inefficient usage of energy by underloaded servers, this might lead to serious QoS degradation issues in the overloaded servers.

In this paper, we propose a lightweight decentralized solution for homogenizing the load across different machines in a data center by mapping the problem to a Stable Marriage matching problem. The algorithm judiciously chooses pairs of overloaded and underloaded servers for matching and subsequently VM migrations are performed to reduce load imbalance. For the purpose of comparisons, three different greedy matching algorithms are also introduced. In order to verify the feasibility of our approach in real-life scenarios, we implement our solution on a small test-bed. For the larger scale scenarios, we provide simulation results that demonstrate the efficiency of the algorithm and its ability to yield a near-optimal solution compared to other algorithms. The results are promising, given the low computational footprint of the algorithm.

Keywords: Self-organization · Cloud computing · Stable Marriage · Distributed load balancing

1 Introduction

Major systems and Internet based services have grown to such a scale that we now use the term "hyper scale" to describe them. Furthermore, hyper scale architectures are often deployed in cloud based environments, which offers a flexible pay-as-you-go model.

© Springer-Verlag GmbH Germany, part of Springer Nature 2020
A. Hameurlain et al. (Eds.) TLDKS XLV, LNCS 12390, pp. 172–198, 2020.
https://doi.org/10.1007/978-3-662-62308-4_7

From a system administrator's perspective, optimizing a hyper scale solution implies introducing system behaviour that can yield automated reactions to changes in configurations and fault occurrences. For instance, auto scaling is a desired behaviour model for websites to optimize cost and performance in accordance to usage patterns.

There are two different perspectives on how an automated behaviour can be implemented within the field of cloud computing. One of the perspectives is to implement the behaviour in the infrastructure, which is the paradigm embraced by the industry. The other alternative is to introduce behaviour as a part of the Virtual Machine (VM), which opens up a possibility for cloud independent models.

Several studies have shown that the resource utilization within a data center varies drastically across the physical servers [4,12,28]. Load imbalance situations are observed where a significant portion of servers are either in overloaded or underloaded states. Apart from leading to inefficient usage of energy by the presence of underloaded servers, this might lead to serious QoS degradation issues in the overloaded servers. The aim of this paper is to present an efficient and yet simple solution for homogenizing the load in data centers. The potential gain with this research is to find an efficient and less complex way of operating a data center. Stable Marriage is a an intriguing theory emanating from the field of economy and holds many promises in the field of computer science and more particularly in the field of cloud management. In this paper, we apply the theory of Stable Marriage matching in order to devise a load homogenizing scheme within a data center. We also modify the original algorithm in order to support distributed execution.

Various studies on self-organizing approaches have been emerging in the recent years to efficiently solve computationally hard problems where centralized solutions might not scale or might also create a single point a failure.

The aim of this paper is to provide a distributed solution for achieving distributed load balancing in a data center which is inspired by the the Stable Marriage algorithm [16]. The algorithm implements message exchange between pairs of servers. It is worth emphasizing that modern distributed systems often use gossip protocols to solve problems that might be difficult to solve in other ways, either because the underlying network has an inconvenient structure, is extremely large, or because gossip solutions are the most efficient ones available [17] in terms of communication. We shall adopt the Stable Marriage algorithm and study its behavior under different scales.

2   Stable Matching

According to Shapley et al. [15] an allocation where no individual perceives any gain from any further trade is called *Stable*. Stability is a central theory in the field of cooperative game theory that emanates from mathematical economics which seeks to know how any constellation of rational individuals might cooperate to choose an allocation.

Shapley [14] introduced the concept of *pairwise matching*. Pairwise matching studies how individuals can be paired up when they all have different preferences regarding who are their best matches. The matching was analyzed at an abstract level where the idea of marriage was used as an illustrative example.

For this experiment Shapley et al. tested how ten women and ten men should be matched, while respecting their individual preferences. The main challenge was to find a simple method that would lead to stable matching, where no couples would break up and form new matches which would make them better off. The solution was *deferred acceptance*, a simple set of rules that always led straight to the stable matching.

Deferred acceptance can be set up in two different ways, either men propose to women or women propose to men. If women propose to men the process begins with each woman proposing to the man she likes the best. Each man then looks at the different proposals he has received, if any, and regards the best proposal and rejects the others. The women who were rejected in the first round, then move along to propose to their second best choice. This will continue in a loop until no women wants to make any further proposals. Shapley et al. [15] proved this algorithm mathematically and showed that this algorithm always leads to stable matching.

The specific way the algorithm was set up turned out to have an important distributional consequence. The outcome of the algorithm might differ significantly depending on whether the right to propose was given to the women or to the men. If men proposed this lead to the worst outcome from the women's perspective. This is because if women proposed, no woman would be worse off than if the men had been given the right to propose [15].

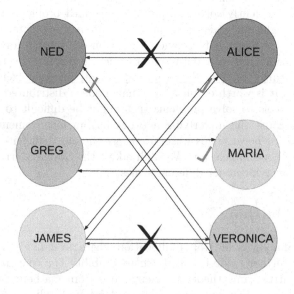

Fig. 1. Stable matching

The model depicted in Fig. 1 presents the selection process for Stable Matching. On the right side we find the women with their preferences and to the left the men with their respective preferences.

3   Related Work

In this section, we shall review some prominent works on distributed approaches for homogenizing the load in a data center. It is worth mentioning that the related work in this particular area is rather sparse.

Marzolla et al. [17] propose a decentralized gossip-based algorithm, called *V-MAN* to address to the issues regarding consolidation. V-MAN is invoked periodically to create consolidate VMs into fewer servers. They assume that the cloud system has a communication layer, so that any pair of servers can exchange messages between them. The work of Marzolla et al. [17] yields very promising results which show that using V-MAN converges fast – after less than 5 rounds of message exchanging between the servers.

In [3], the authors use scouts which are allowed to move from one PM (physical machine) to another – to be able to recognize which compute node might be a suitable migration destination for a VM. This is completely opposite of what V-MAN does. It does not rely on any subset like scouts, instead each server can individually cooperate to identify a new VM location, which makes V-MAN scalable. It is also to be noted that any server can leave or join the cloud at any time.

Sedaghat et al. [25] use a Peer-to-Peer protocol to achieve energy efficiency and increase the resource utilization. The Peer-to-Peer protocol provides mechanisms for nodes to join, leave, publish or search for a resource-object in the overlay or network. This solution also considers multi-dimensionality – because the algorithm needs to be specified to be dimension aware, each PMs proportionality should be considered. Each node is a peer where a peer sampling service, known as newscast, provides each peer with a list of peers whom are to be considered neighbours. Each peer only know k random neighbours which map its local view. The aim is to improve a common value which is defined as the total imbalance of each pair at the time of decision-making by redistributing the VMs. The work uses a modified dimension aware algorithm to tackle the multi-dimensional problem. The algorithm is iterative and starts from an arbitrary VM placement. When the algorithm converges, a reconfiguration plan is set so the migration of the VMs can start.

A survey by Hummaida et al. [10] is focused on adaptation of computing resources. In [29], a peer-to-peer distributed and decentralized approach is proposed that enables servers to communicate with each others without a centralized controller. It uses a node discovery service which is run periodically to find new neighbouring servers to communicate with. The algorithm decides whether two servers should exchange a VM based on the predefined objectives. Similarly, in [20], the authors propose a decentralized approach for user-centric elasticity management to achieve conflict free solutions between customer satisfaction and business objectives. Dynamical allocation of CPU resources to servers is performed in [13] which integrates the Kalman filter into feedback controllers to track the CPU utilizations and update the allocations accordingly.

Siebenhaar et al. [26] use a decentralized approach to achieve better resource management using a two phase negotiation protocol to conduct negotiations with multiple providers across multiple tiers. Moreover, in [8] the authors present an architecture for dynamic resource provisioning via distributed decisions where each server makes its own utilization decision based on its own current capacity and workload characteristics. The authors also introduce a light-weight provisioning optimizer with a replaceable routing algorithm for resource provisioning and scaling. The authors claim that with this solution the resource provisioning system will be more scalable, reliable, traceable, and simple to manage.

Calcavecchia et al. [6] start by criticizing centralized solutions and states that it is not suitable for managing large-scale systems. The authors introduce a Decentralized Probabilistic Auto-Scaling Algorithm (DEPAS) which is integrated into a P2P architecture and allows simultaneous auto-scaling of services over multiple cloud infrastructures. They conduct simulations and real platform based experiments and claims that their approach is capable of handling (i): keeping track of: overall utilization of all Instant Cloud resources within the target range, (ii): maintaining service response times close to those achieved through optimal centralized auto-scale approaches.

Another interesting decentralized approach is presented by [5] where an emphasis is given to low price based deployment and dynamic scaling of component based applications to meet SLA performance and availability goals. The work gives priority to a low price model instead of normalizing loads to all servers, meaning the dynamic economic fitness of the servers will decide whether resources are replicated, migrated to another server, or deleted. Each server stores the table of complete mapping between instances and servers and a gossip protocol is used for mapping instances.

In another survey [21] the authors aim to classify and provide a concise summary of the several proposals for cloud resource elasticity today. They present a taxonomy covering a wide range of aspects, and discuss details for each of the aspects, and the main research challenges. Finally, they propose fields that require further research. A more recent article on cloud computing elasticity [1] reviews both classical and recent elasticity solutions and provides an overview of containerization. It also discusses major issues and research challenges related to elasticity in cloud computing. The authors comprehensively review and analyze the proposals developed in this field.

In a survey on elasticity management in PaaS systems [19] the authors claim that ideally, that elasticity management should be done by specialised companies: the platform as a service (PaaS) providers. The PaaS layer is placed on top of an IaaS layer in a regular cloud computing architecture. The authors provide a tutorial on the requirements to be considered and the current solutions to the challenges being present in elastic PaaS systems and conclude that elasticity management in the PaaS service model is an active research area amenable to improvement.

Most of the papers mentioned above concern scaling of resources and load distribution by developers. However, there has been some recent development in the field of cloud computing which might entirely shift the burden of scaling resources from developers and system designer to cloud providers. This new cloud computing paradigm is called Serverless Computing. The term Serverless Computing is a platform (Function-as-a-service) that hides the server usage from the developer and runs code on-demand, scaling accordingly and only billing for the time the code is running [7].

The term Serverless can however be a bit misleading, serverless doesn't equal to no backend service, it just means that all of the server space and infrastructure issues are handled by the vendor [27]. Many larger companies (e.g Amazon AWS, Azure, etc.) now offer these type of solutions, where you can deploy your code to the cloud, and only pay for the amount of time the code is used, while all the administration of the servers are handled by the vendor [2].

AWS was one of the first vendors to introduce the concept of serverless computing in 2014 [11]. According to Castro et al. [7] serverless seems to be the natural progression in the advancement and adoption of VM and container technology, where each layer of the abstraction leads to more lightweight units of computation, saving resources, being more cost effective and increasing the speed of development. Castro et al. conclude that serverless computing lowers the bar for developers by delegating to the platform provider much of the operational complexity of monitoring and scaling large-scale applications. However, the developer now needs to work around limitations on the stateless nature of their functions, and understand how to map their application's SLAs to those of the serverless platform and other dependent services [7].

4 Solution

4.1 Overview of a Functioning Framework

As the algorithm implemented will be based on a real life inspiration, it is important to understand that the outcome can end in two different cases. Just as each relationship does not end in marriage neither will the decision of the PMs. Each PM can be viewed as individuals making their own "life choices".

Figure 2 and 3 enhances the different outcomes the algorithm can opt for and how the framework is set up to work around the execution of the algorithm. Note that the environment later implemented is not in an actual data center. Our main goal is to achieve load balance in a distributed cloud data center, but our solution is restricted to load-balancing of CPU-intensive applications as we have not considered the impact of other resources such as memory, network and disk. The intention is to create a framework that can handle any given scenario or setup for CPU-intensive applications.

The basic framework for both scenarios are the same, it is a data center consisting of PMs with different weight. However, as explained in the section above, based on the calculations of the underloaded server in the second scenario the proposal is rejected and the PM moves on to the next best on their list. This

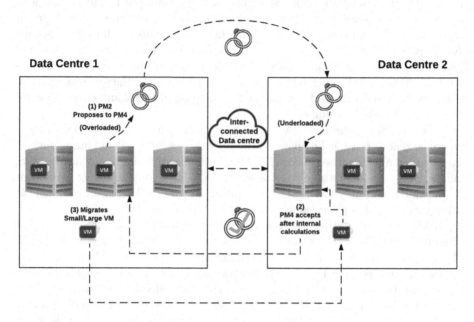

Fig. 2. Proposal accepted

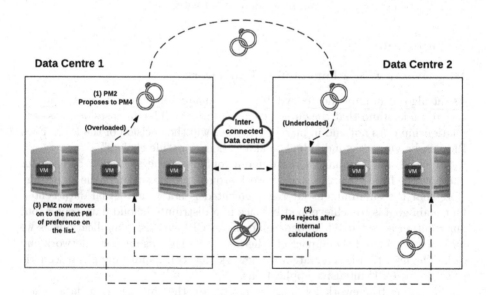

Fig. 3. Proposal rejected

process is supposed to be a continuous process, unless the target load for each PM is achieved, then the process stops entirely.

4.2   Bin Packing with Stable Marriage

The bin packing problem is the challenge of packing an amount n of objects in to as few bins as possible. In this case, the servers are the bins and the VMs are the objects. This terminology fits the Stable Marriage algorithm well, as the bins are the humans who are in quest of a partner (bin) which represents a good match while satisfying some constraints in terms of capacity. A constraint can be defined in many ways, but for a bin some common constraints would be the height of the box, its width and depth. Our algorithm will focus on Virtual CPUs (VCPUs) and memory as constraints. The aim of the algorithm is to even and equalize the load of the data center by evenly distribute the weight of the VMs across the bins in such a way that the bins should neither be overfilled nor underfilled.

4.3   Stable Marriage Animation

A known set of servers is divided into two groups of overutilized (men) and underutilized (women). The goal is to find a perfect match for the overutilized servers. The matchmaking is based on three values, the average CPU load, the imbalance before and after migration (calculated before the eventual migration) and the profit of such a marriage.

The figures below demonstrate the expected outcome of implementing the Stable Marriage algorithm. This approach is mainly centralized and the PMs know the allocated values of each other. This means that each PM, both over and underutilized, has a list of preferred men and women they want to propose to or receive a proposal from.

Figure 4 shows that PM1 has reached full capacity as marked by the red line. The red line represents the average capacity that each PM can handle. Assume that each group of men can only handle four or six full servers, in this case PM1 has then reached its full capacity and so has PM2. They need to migrate the load to a underloaded PM of preference, so that they can balance the load equally. Hence, the overloaded server PM1 proposes to his first choice, PM3.

The female set of servers have their own method to calculate the advantage/disadvantage of such a marriage. If the underutilized PM calculates a higher imbalance than before the marriage, she sees this as a disadvantage and rejects the proposal. This method also avoids that the proposing party becomes underutilized in the future.

After being rejected, PM1 proceeds to his next best choice which is PM4. PM4 then calculates the imbalance before and after the proposed marriage to check if it improves after a potential migration. In this case the imbalance factor improves, and PM4 accepts the proposal. The migration can now take place.

Since PM4 has the same amount of capacity to accept load, the server is not over-utilized and the load has been balanced between the married PMs.

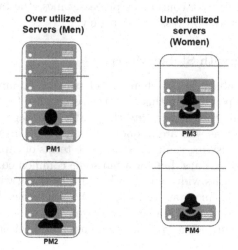

Fig. 4. Set of over/under utilized servers

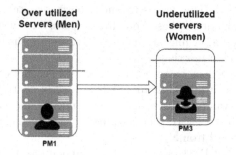

Fig. 5. PM1 proposes to PM3

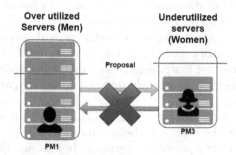

Fig. 6. PM3 rejects PM1 seeing no benefit to this marriage.

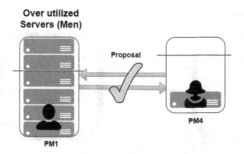

Fig. 7. PM4 accepts PM1's proposal

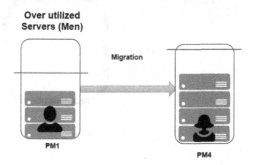

Fig. 8. Migration successful

This particular animation doesn't have any scheme implemented, it just gives an idea of how the algorithm is supposed to work. The schemes will only make a difference in terms of the size of the VMs that is migrated. The figure below shows how the VM sizes may differ on each PM and how the migration process may look inside each server.

4.4 The Proposed Stable Marriage Algorithm

In our preliminary version of this work [24], we proposed two implementations of the Stable Marriage algorithm called *Migrate Smallest-Greedy Matching* (MS-GM) and *Migrate Largest-Greedy Matching* (ML-GM). We observed that both of the methods have some pitfalls. The *Migrate Smallest* algorithm usually results in a better overall result, but at the cost of a higher number of migrations. On the other hand, the *Migrate Largest* algorithm requires a smaller number of migrations in order to reach a final state at the cost of a higher final imbalance. In order to improve on these two algorithms, we propose a Stable Marriage based approach which will provide a minimum imbalance result while still yielding a low number of migrations. In addition to this, in order to compare with the efficiency of new the Stable Marriage (SM) algorithm, three different kinds of greedy matching algorithms are introduced.

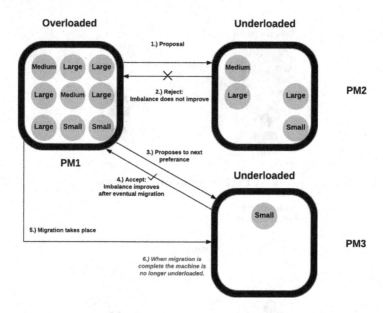

Fig. 9. PMs with various VM sizes

In the example shown in Fig. 9 PM1 is overloaded while PM2 and PM3 are underloaded. As can be seen, PM1 is highly overloaded and PM3 is highly underloaded, therefore the SM algorithm pairs PM1 and PM3 and migrates the largest VM of PM1 to PM3. Then the algorithm will calculate the imbalance of each PM again and compute the overload and underload. Based on the amount of overload and underload, SM will determine a new pair of PMs and move the optimal VM based on size. This process will continue until a minimum imbalance is achieved. Figure 10 illustrates that the resources allocated to the VMs are different. The resources of the three VMs correspond to those of the three smallest E2 high-CPU machine types of Google Cloud Engine. We restrict our study to these three VM sizes as it simplifies the solution while this design choice still is complex enough to illustrate the usefulness of our solution. It is straightforward to generalize this by simply including VMs of different sizes. It should be noted that in the experiments only the number of CPUs is taken into account when determining whether a server is underloaded or overloaded. In a more extensive solution other resources and features, like memory, network bandwidth, disk usage, runtime monitoring, message overheads and migration time should be taken into account. However, for CPU intensive applications our solution includes the most important feature.

The complexity per time instance of the SM algorithm is $O(n_o n_u)$ where n_o and n_u are the numbers of overloaded, respectively underloaded servers at each time instant. Thus, the number of messages exchanged is in the order of $O(n_o n_u)$ at each iteration of the algorithm. The complexity of the greedy ones is $O(n_u)$ since the overloaded servers are those that propose. The upper bound on the running time (and the number of migrations) is the total number of VMs in all servers.

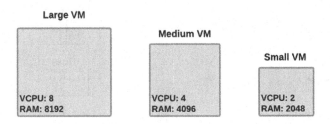

Fig. 10. VMs with their allocated resources.

4.5   The Greedy Matching Approach

The greedy matching approach is inspired by the greedy algorithm. In this approach we first take the one most overloaded server and compare the imbalance with every underloaded server with a predefined movement method and choose the one that will reduce the imbalance most. If there are multiple cases which have the same imbalance result, one of them will be picked at random. Once this process is done, the algorithm again calculates the imbalance of each server and continues the above process until the minimum imbalance result is gained. The three predefined movement methods are denoted as Migrate Smallest (MS), Migrate Medium (MF) and Migrate Largest (ML). For each of the methods, a VM of the size corresponding to the name of the method is migrated from the most overloaded PM to the most underloaded PM.

5   Implementation of Stable Marriage

In this section we give details about the implementation of the Stable Marriage algorithm which aims to find the perfect match for gaining load balance. Each node is considered an individual with preferences and demands. These are taken into consideration to be able to find the perfect balance for each individual node.

The flow diagram in Fig. 11 gives an insight into how the Stable Marriage Algorithm operates and the different procedures involved.

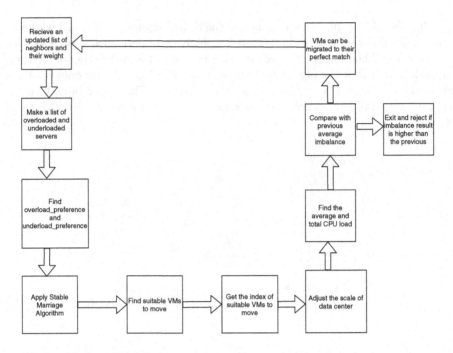

Fig. 11. Flow diagram of the stable marriage implementation

The following describes the most important parameters of the algorithm inspired by Rao et al. [23]. We define the CPU load to simply be the number of VCPUs of a VM. Let C_i be the total CPU load of server i contained in its VMs and let c_j be the number of VCPUs assigned to VM_j:

$$C_i = \sum_{VM_j} c_j$$

Each server i has a maximum CPU capacity $C_i{}^{max}$ which is its number of physical CPUs and we define this to be the maximum number of VCPUs a server can contain:

$$C_i \leq C_i{}^{max}$$

When it comes to consolidation, most algorithms take into account the bottleneck resource as a sole criterion for achieving better consolidation decisions. Similarly, when it comes to load balancing, one can base the algorithm on the imbalanced resource whether it is CPU or memory. For the sake of simplicity, we assume that the CPU is the most imbalanced resource in our data center, which in real life is often the case.

Average load $\overline{C}$ is defined as the average CPU load of the N Physical Machines:

$$\overline{C} = \sum_{PM_i} C_i/N = \sum_{i=1}^{N} C_i/N$$

If the system was perfectly balanced and all servers of same size, each server would have this number of VCPUs. Furthermore, we define the average capacity of a server as

$$\overline{C^{max}} = \sum_{PM_i} C_i{}^{max}/N$$

We define the target load to be the result when evenly distributing the load according to the capacity of the servers. Let T_i be the the target load at PM_i when there is no imbalance:

$$T_i = \frac{C_i{}^{max}}{\overline{C^{max}}}\overline{C}$$

If all the machines have the same capacity, this would reduce to equal load on each server:

$$T_i = \overline{C} = \sum_{PM_i} C_i/N$$

We define the imbalance I_i of a server or physical machine PM_i in terms of CPU load as the deviation of the load of machine PM_i from the target CPU load:

The following pseudo code shows how the possible gain of a migration between an overloaded server and an underloaded server is calculated:

```
Gain_of_Migration_Couple
# Calculate imbalance before an eventual migration
<calculate imbalance of overloaded server>
<calculate imbalance of underloaded server>
overloaded_pref = [make preferences based on overloaded size ]
underloaded_pref = [make preferences based of underloaded size]
SM = stablemarriage(overloaded_pref, underloaded_pref)
for each pair on SM
    check difference and move suitable VMs
continue the process until minimum imbalance is achieved
```

The Stable Marriage algorithm operates in rounds and it stops when no more "gain" in terms of reducing imbalance can be achieved. The algorithm will then exit. In other terms, if there is no beneficial proposal that reduces the imbalance or the proposals will increase the imbalance, the algorithm will stop whenever there are no possibilities to reduce further imbalance. It also restricts overloaded servers to become underloaded, which means that PMs may also decline a proposal if the overloaded server becomes underloaded. This means that a node can never become overbalanced again or underbalanced to take more VMs on board. This is an important part of the implementation, as the point of the Stable Marriage algorithm is to *stabilize* the system, this algorithm contributes to the stability factor.

6   Experiments

6.1   Experimental Set-Up

Figure 12 is a model which gives an overview of the structure in which the project will be implemented. This is a figure which shows how the different components from entirely different worlds are paired together. The bottom layer is the physical hardware consisting of PM1-PM3 or Lab01-Lab03 which are the assigned name on the OS. This layer is controlled by the hypervisor KVM, which is in control of the virtual environment, also the network of VMs which are later spawned in layer 3.

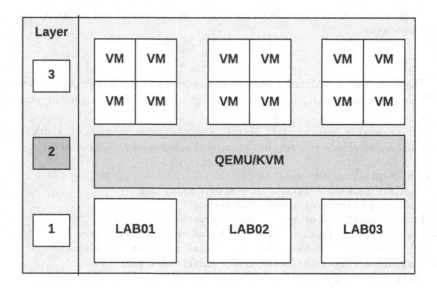

Fig. 12. Design

6.2   Environment Configuration

Evidently, a framework is built with several services and components, which are necessary for an environment to work. To set up a virtual environment for this project several physical and virtual technologies were necessary.

   The physical servers in this project are stored in a server room at Oslo Metropolitan University. There are three dedicated servers for this project, as seen in Fig. 13. The setup consists of a dedicated gateway to connect to the outside. All of the PMs are inter-connected through a dedicated switch.

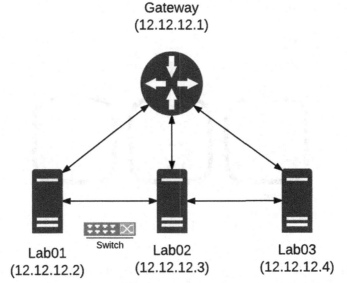

Fig. 13. Overview of the physical lab structure

Each server is allocated with same specifications (Fig. 14):

Hardware:	Design:
Processor	Intel(R) Core(TM)2 Duo CPU E7500 @ 2.93GHz
Architecture	x86_64
Memory	2048 MB
CPU	2
Operating System	Ubuntu 16.04.2 LTS (Xenial)
Virtual	QEMU/KVM, Libvirt

Fig. 14. Physical attributes

These are the details for the physical hardware which are dedicated for the virtual implementation. The PMs run Ubuntu which is easier to work with especially with QEMU and KVM for virtualization of the environment.

6.3   Virtual Configuration

The next step is to configure the virtual network. This network will also ensure that when migrating VMs from one host to another, this happens within the same virtualized network. Figure 15 below shows how the PMs are connected and how the VMs reside inside the PMs. The VMs are attached to a virtual bridge by birth. This is a actually a virtual switch, however it is called a bridge and used with KVM/QEMU hypervisors to be able to use live migration for instance.

To connect the PMs together, a physical switch is used.

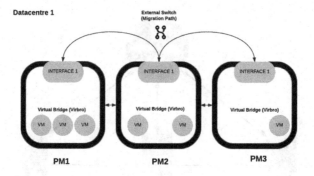

Fig. 15. Physical lab details

The different VM flavors that will be used in the experiments are given in Fig. 16. Each PM will be given a combination of VMs of different flavors.

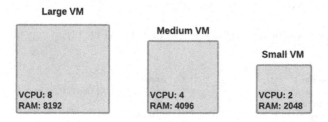

Fig. 16. VM flavors

6.4 Comparison of Greedy Matching and Stable Marriage

The aim of these experiments is to show and compare the impact of different types of greedy matching approaches with the Stable Marriage migration approach in terms of imbalance and the number of migrations taking place in a virtualized environment with different flavours of VMs. For the sake of clarity, all the abbreviations used in the algorithms are summarized in the Table 1. To test the effectiveness of the migration algorithms, three different test scenarios have been created, TEST-10, TEST-50 and TEST-100. The TEST-10 experiment refers to a small scale system with a bin size of 10 and where each bin or PM contains a number of VMs between 5 and 10. Similarly, TEST-50 refers to medium system with bins size of 50 where each bin contains a random number of VMs in the range 40 to 50. Lastly, TEST-100 refers to a large scale system, with 100 bins and each bin has a random number of VMs between 80 and 100. The results presented below are the average result of 100 experiments. In order to find the most effective approach each experiment is performed using all the four algorithms.

Table 1. List of abbreviations used for algorithms

Abbreviation	Definition
GM	Greedy Matching
SM	Stable Marriage
MS-GM	Migrate Smallest-Greedy Matching
MM-GM	Migrate Medium-Greedy Matching
ML-GM	Migrate Largest-Greedy Matching

Migrate Smallest-Greedy Matching vs Stable Marriage. Figure 17 and
18 depict the results for the three different test scenarios and shows the comparison between Migrate Smallest-Greedy Matching (MS-GM) and Stable Marriage (SM)
migration approaches in terms of imbalance and migrations. In TEST-10, the initial
average imbalance was 8.65 which is reduced to 1.18 by the Migrate Smallest-Greedy
Matching (MS-GM) approach while the Stable Marriage approach minimized the initial
imbalance to 0.83. Interestingly in Fig. 18, we can see the Stable Marriage approach
takes less than half the number of VM migrations compared to the MS-GM approach
for the two largest experiments.

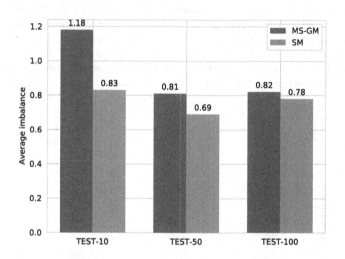

Fig. 17. Migrate Smallest-Greedy Matching (MS-GM) vs Stable Matching (SM)

In TEST-50, the initial average imbalance was 18.06. The MS-GM approach
brought down this imbalance to 0.81 after 399 VM migrations while SM reduced the
initial average imbalance to 0.69 after only 145 VM migrations. In TEST-100, the initial average imbalance was 29.82. The SM approach required 223 migrations to reduce
the imbalance to 0.78 while MS-GM required more than three times this number of
migrations, 704, and minimized the imbalance to 0.82 which is higher than the result
of SM.

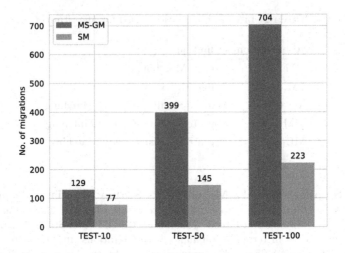

Fig. 18. MS-GM migrations vs SM migrations

Migrate Medium-Greedy Matching vs Stable Marriage. Figure 19 shows a comparison of the results of Migrate Medium-Greedy Matching and Stable Marriage approach obtained from three different tests. In TEST-10, the MM-GM approach reduced the initial average imbalance to 2.59 while the SM approach yielded a more than three times smaller imbalance result. From Fig. 20 we see that SM not only gave better imbalance results but it also needed a smaller number of migrations. Also for TEST-50 and TEST-100, SM provided better imbalance results compared to MM and needed fewer migrations in order to reach these results.

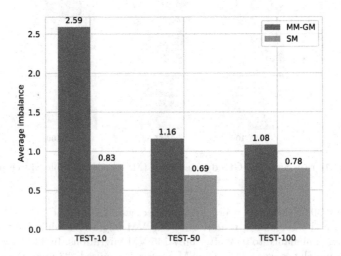

Fig. 19. Migrate Medium-Greedy Matching (MM-GM) vs Stable Matching (SM)

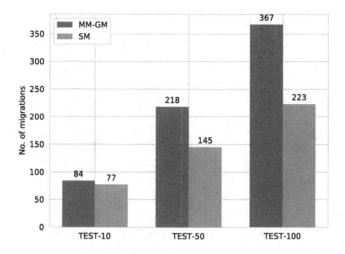

Fig. 20. MM-GM migrations vs SM migrations

Migrate Largest-Greedy Matching vs Stable Marriage. Figure 21 shows the results obtained from the three different tests for Migrate Largest-Greedy Matching (ML-GM) and Stable Marriage (SM). In this comparison, we can observe that in all the three tests, ML was not able to reduce the initial average imbalance to less than 2 while SM minimized the imbalance to less than 0.83 all cases. On the other hand, in Fig. 22 we see that ML-GM required slightly fewer VM migrations than SM. This is because when ML-GM migrates VMs it always moves the largest VMs and then the total number of migrations will be smaller than for other methods.

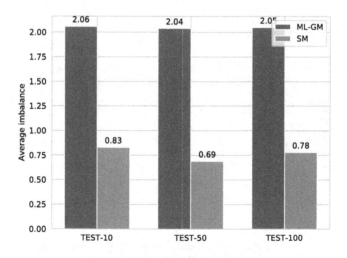

Fig. 21. Migrate Largest-Greedy Matching (ML-GM) vs Stable Matching (SM)

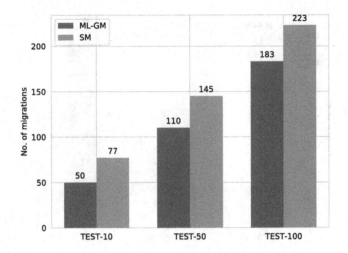

Fig. 22. ML-GM migrations vs SM migrations

In TEST-100, the ML-GM approach reduced the initial average imbalance to 2.05 from 29.82 whereas SM minimized it to 0.78 which is a nearly three times better result. However, ML-GM needed just 183 migrations and SM needed 41 more migrations to reach its lowest imbalance state.

Overall Imbalance Reduction. Figure 23 illustrates how the average imbalance was reduced as function of the number of migrations for the four algorithms tested for the TEST-100 experiment. For the greedy matching algorithms, as the size of the VMs

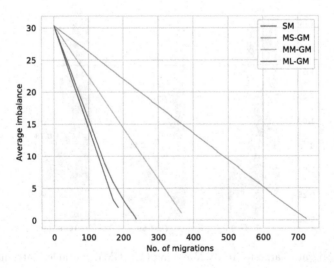

Fig. 23. Imbalance reduction as function of number of migrations in the TEST-100 experiment.

moved decreases an increasing number of iterations are needed to reach the final state. This is not surprising as a larger number of small VMs must be moved when reducing an imbalance of the same magnitude compared to when moving large VMs. However, migrating using small VMs leads to a final state with smaller imbalance. The final state of the SM algorithm has on average the smallest imbalance and at the same time it uses just a few more migrations to reach the final state compared to the ML-GM algorithm.

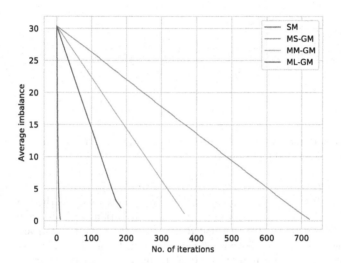

Fig. 24. Imbalance reduction as function of number of iterations in the TEST-100 experiment.

Figure 24 shows the reduction of average imbalance as function of the number of iterations needed to reach the final state of smallest possible imbalance. It is apparent that the SM algorithm needs an order of magnitude fewer iterations to find imbalance compared to the greedy algorithms. This is for a large part due to the fact that the SM algorithm for each iteration migrates VMs between all the bins in parallel while the GM algorithms migrates just a single VM for each iteration. Nevertheless, this means that the SM algorithm in a real data center will reach a balanced state much faster than the other algorithms.

Figure 25 shows the average imbalance reduction of our proposed SM algorithm as function of number of iterations. In the first few iterations the reduction ratio is very high while for the last few iterations the reduction in average imbalance is very small as the optimal balance is reached.

194   D. Sangar et al.

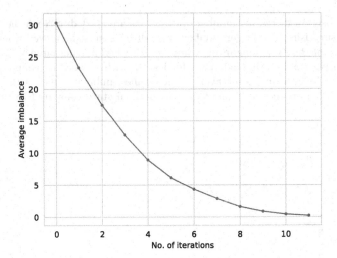

Fig. 25. Reduction of imbalance as function of iterations for the Stable Marriage algorithm in the TEST-100 experiment.

Figure 26 illustrates how the number of migrations varies as the iterations of our proposed SM algorithm is performed. In the first iteration, 50 VMs are migrated from 50 overloaded servers to 50 underloaded servers at the same time, meaning that all of the 100 bins of the TEST-100 experiment take part. In the second iteration 40 parallel migrations take place, meaning that 20 of the bins have already reach a state close to be balanced. The SM algorithm continues in the same manner efficiently reducing the imbalance and reaches a balanced state after only 11 iterations. On the other hand,

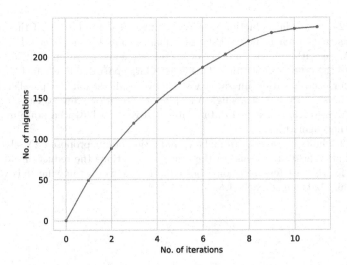

Fig. 26. Number of migrations as function of number of iterations for the Stable Marriage algorithm in the TEST-100 experiment.

the greedy matching algorithms migrates just a single VM for each iteration and thus spend much more time reaching a balanced situation.

Performance Optimization. In the case of load management, the optimal solution is to have a perfect load balance between servers, i.e., to reduce the imbalance to 0. The load distribution problem is known to be NP hard and hence intractable for large systems [9,18]. Our goal is to design a system that yields near-optimal solutions.

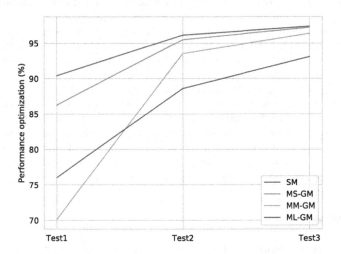

Fig. 27. Performance optimization of the proposed Stable Marriage and greedy matching algorithms.

Figure 27 illustrates the performance optimization of the proposed stable marriage and three different greedy matching algorithms in terms of imbalance reduction in three different test cases. In all three test scenarios the proposed stable marriage algorithm is clearly best, while ML-GM and MM-GM have the worst optimization percentage. However, in the case of test3, the MS-GM performance is close to the one of the SM algorithm. But as we already presented in Fig. 23 and Fig. 24, SM-GM requires a larger number of migrations and iterations which leads to extra overhead in terms of processing and time.

In a recent paper dealing with load balanced task scheduling for cloud computing, Panda et al. [22] propose a new algorithm reckoned as Probabilistic approach for Load Balancing (PLB). They run extensive simulations and compare their results with several algorithms which have been used in this context namely, Random, Cloud List Scheduling (CLS), Greedy and Round Robin (RR) [30]. The problem of load balancing tasks on VMs is comparable to load balancing VMs on physical servers and one of their Key Performance Indicators (KPI) is the dispersion of a set of loads from its average load, which is comparable to the load imbalance metric reported above [23]. In the simulations of Panda et al. the results of the Greedy algorithm are roughly equal to the performance of the other classical algorithms. We see this as an indication of how well our SM algorithm performs compared to other algorithms traditionally used for load balancing in cloud computing.

7 Conclusion

In our preliminary work [24], we addressed the problem of homogenizing the load in a cloud data center using the concept of the Stable Marriage algorithm. The results were promising and demonstrated the ability of the proposed algorithms to efficiently distribute the load across different physical servers. Although the proposed algorithms in [24] are based on the principles of Stable Marriage, they have a greedy matching principle and fail to replicate and incarnate the ideas of Stable Marriage theory in a proper way. Furthermore, in the analysis of results we found that the Migrate Smallest method gives better imbalance results but requires a large number of migrations while Migrate Smallest requires smaller number of migrations but at the cost of a larger final imbalance. To overcome this problem we propose in this paper a more conform Stable Marriage algorithm and compare its results to three different types of greedy matching algorithms. Based on the results presented in Sect. 6.4, we can conclude that the Stable Marriage algorithm yields the lowest imbalance result combined with the lowest number of migrations. A single exception is that the Migrate Largest algorithm needs slightly fewer migrations, but its final imbalance is then larger. The Stable Marriage algorithm requires three times fewer migrations and provides the lowest imbalance compared to the other two greedy algorithms. Our results also show that the number of iterations needed to reach a balanced state is an order of magnitude smaller than the number of iterations needed by the three greedy matching algorithms.

References

1. Al-Dhuraibi, Y., Paraiso, F., Djarallah, N., Merle, P.: Elasticity in cloud computing: state of the art and research challenges. IEEE Trans. Serv. Comput. **11**(2), 430–447 (2017)
2. Amazon: Serverless computing (2020). https://aws.amazon.com/serverless/. Accessed 17 June 2020
3. Barbagallo, D., Di Nitto, E., Dubois, D.J., Mirandola, R.: A bio-inspired algorithm for energy optimization in a self-organizing data center. In: Weyns, D., Malek, S., de Lemos, R., Andersson, J. (eds.) SOAR 2009. LNCS, vol. 6090, pp. 127–151. Springer, Heidelberg (2010). https://doi.org/10.1007/978-3-642-14412-7_7
4. Barroso, L.A., Hölzle, U., Ranganathan, P.: The datacenter as a computer: designing warehouse-scale machines. Synth. Lect. Comput. Archit. **13**(3), i-189 (2018)
5. Bonvin, N., Papaioannou, T.G., Aberer, K.: Autonomic SLA-driven provisioning for cloud applications. In: 2011 11th IEEE/ACM International Symposium on Cluster, Cloud and Grid Computing, pp. 434–443. IEEE (2011)
6. Calcavecchia, N.M., Caprarescu, B.A., Di Nitto, E., Dubois, D.J., Petcu, D.: DEPAS: a decentralized probabilistic algorithm for auto-scaling. Computing **94**(8–10), 701–730 (2012)
7. Castro, P., Ishakian, V., Muthusamy, V., Slominski, A.: The rise of serverless computing. Commun. ACM **62**(12), 44–54 (2019)
8. Chieu, T.C., Chan, H.: Dynamic resource allocation via distributed decisions in cloud environment. In: 2011 IEEE 8th International Conference on e-Business Engineering, pp. 125–130. IEEE (2011)
9. Garey, M.R., Johnson, D.S.: Computers and Intractability, vol. 174. Freeman, San Francisco (1979)

10. Hummaida, A.R., Paton, N.W., Sakellariou, R.: Adaptation in cloud resource configuration: a survey. J. Cloud Comput. **5**(1), 1–16 (2016). https://doi.org/10.1186/s13677-016-0057-9
11. Jangda, A., Pinckney, D., Brun, Y., Guha, A.: Formal foundations of serverless computing. In: Proceedings of the ACM on Programming Languages 3 (OOPSLA), pp. 1–26 (2019)
12. Jin, C., Bai, X., Yang, C., Mao, W., Xu, X.: A review of power consumption models of servers in data centers. Appl. Energy **265**, 114806 (2020)
13. Kalyvianaki, E., Charalambous, T., Hand, S.: Self-adaptive and self-configured CPU resource provisioning for virtualized servers using Kalman filters. In: Proceedings of the 6th International Conference on Autonomic Computing, pp. 117–126 (2009)
14. Levine, D.K.: Introduction to the special issue in honor of Lloyd Shapley: eight topics in game theory. Games Econ. Behav. **108**, 1–12 (2018). https://doi.org/10.1016/j.geb.2018.05.001. http://www.sciencedirect.com/science/article/pii/S089982561830068X. Special Issue in Honor of Lloyd Shapley: Seven Topics in Game Theory
15. Lloyd Shapley, A.R.: Stable matching: theory, evidence, and practical design. https://www.nobelprize.org/uploads/2018/06/popular-economicsciences2012.pdf
16. Manlove, D.F.: Algorithmics of Matching Under Preferences, vol. 2. World Scientific, Singapore (2013)
17. Marzolla, M., Babaoglu, O., Panzieri, F.: Server consolidation in clouds through gossiping. In: 2011 IEEE International Symposium on a World of Wireless, Mobile and Multimedia Networks (WoWMoM), pp. 1–6. IEEE (2011)
18. Mishra, S.K., Sahoo, B., Parida, P.P.: Load balancing in cloud computing: a big picture. J. King Saud Univ. Comput. Inf. Sci. **32**(2), 149–158 (2020)
19. Muñoz-Escoí, F.D., Bernabéu-Aubán, J.M.: A survey on elasticity management in PaaS systems. Computing **99**(7), 617–656 (2017)
20. Najjar, A., Serpaggi, X., Gravier, C., Boissier, O.: Multi-agent negotiation for user-centric elasticity management in the cloud. In: 2013 IEEE/ACM 6th International Conference on Utility and Cloud Computing, pp. 357–362. IEEE (2013)
21. Naskos, A., Gounaris, A., Sioutas, S.: Cloud elasticity: a survey. In: Karydis, I., Sioutas, S., Triantafillou, P., Tsoumakos, D. (eds.) ALGOCLOUD 2015. LNCS, vol. 9511, pp. 151–167. Springer, Cham (2016). https://doi.org/10.1007/978-3-319-29919-8_12
22. Panda, S.K., Jana, P.K.: Load balanced task scheduling for cloud computing: a probabilistic approach. Knowl. Inf. Syst. **61**(3), 1607–1631 (2019)
23. Rao, A., Lakshminarayanan, K., Surana, S., Karp, R., Stoica, I.: Load balancing in structured P2P systems. In: Kaashoek, M.F., Stoica, I. (eds.) IPTPS 2003. LNCS, vol. 2735, pp. 68–79. Springer, Heidelberg (2003). https://doi.org/10.1007/978-3-540-45172-3_6
24. Sangar, D., Haugerud, H., Yazidi, A., Begnum, K.: A decentralized approach for homogenizing load distribution: in cloud data center based on stable marriage matching. In: Proceedings of the 11th International Conference on Management of Digital EcoSystems, pp. 292–299 (2019)
25. Sedaghat, M., Hernández-Rodriguez, F., Elmroth, E., Girdzijauskas, S.: Divide the task, multiply the outcome: cooperative VM consolidation. In: 2014 IEEE 6th International Conference on Cloud Computing Technology and Science (CloudCom), pp. 300–305. IEEE (2014)

26. Siebenhaar, M., Nguyen, T.A.B., Lampe, U., Schuller, D., Steinmetz, R.: Concurrent negotiations in cloud-based systems. In: Vanmechelen, K., Altmann, J., Rana, O.F. (eds.) GECON 2011. LNCS, vol. 7150, pp. 17–31. Springer, Heidelberg (2012). https://doi.org/10.1007/978-3-642-28675-9_2

27. Taibi, D., El Ioini, N., Pahl, C., Niederkofler, J.R.S.: Serverless cloud computing (function-as-a-service) patterns: a multivocal literature review. In: Proceedings of the 10th International Conference on Cloud Computing and Services Science (CLOSER 2020) (2020)

28. Vasques, T.L., Moura, P., de Almeida, A.: A review on energy efficiency and demand response with focus on small and medium data centers. Energy Effic. **12**(5), 1399–1428 (2018). https://doi.org/10.1007/s12053-018-9753-2

29. Wuhib, F., Stadler, R., Lindgren, H.: Dynamic resource allocation with management objectives-implementation for an openstack cloud. In: 2012 8th International Conference on Network and Service Management (CNSM) and 2012 Workshop on Systems Virtualiztion Management (SVM), pp. 309–315. IEEE (2012)

30. Xu, M., Tian, W., Buyya, R.: A survey on load balancing algorithms for virtual machines placement in cloud computing. Concurr. Comput. Pract. Exp. **29**(12), e4123 (2017)

A Sentiment Analysis Software Framework for the Support of Business Information Architecture in the Tourist Sector

Javier Murga[1], Gianpierre Zapata[2], Heyul Chavez[3], Carlos Raymundo[4(✉)],
Luis Rivera[5], Francisco Domínguez[5], Javier M. Moguerza[5],
and José María Álvarez[6]

[1] Ingeniería de Software,
Universidad Peruana de Ciencias Aplicadas, Lima, Peru
u201111811@upc.edu.pe
[2] Ingeniería de Sistemas de Información,
Universidad Peruana de Ciencias Aplicadas, Lima, Peru
u201214895@upc.edu.pe
[3] Ingeniería de Telecomunicaciones y Redes,
Universidad Peruana de Ciencias Aplicadas, Lima, Peru
u812426@upc.edu.pe
[4] Dirección de Investigación,
Universidad Peruana de Ciencias Aplicadas, Lima, Peru
carlos.raymundo@upc.edu.pe
[5] Escuela Superior de Ingeniería Informática,
Universidad Rey Juan Carlos, Mostoles, Madrid, Spain
lm.rivera@alumnos.urjc.es, {francisco.dominguez,javier.moguerza}@urjc.es
[6] Department of Computer Science and Engineering,
Universidad Carlos III, Madrid, Spain
joalvare@inf.uc3m.es

Abstract. In recent years, the increased use of digital tools within the Peruvian tourism industry has created a corresponding increase in revenues. However, both factors have caused increased competition in the sector that in turn puts pressure on small and medium enterprises' (SME) revenues and profitability. This study aims to apply neural network based sentiment analysis on social networks to generate a new information search channel that provides a global understanding of user trends and preferences in the tourism sector. A working data-analysis framework will be developed and integrated with tools from the cloud to allow a visual assessment of high probability outcomes based on historical data, to help SMEs estimate the number of tourists arriving and places they want to visit, so that they can generate desirable travel packages in advance, reduce logistics costs, increase sales, and ultimately improve both quality and precision of customer service.

© Springer-Verlag GmbH Germany, part of Springer Nature 2020
A. Hameurlain et al. (Eds.) TLDKS XLV, LNCS 12390, pp. 199–219, 2020.
https://doi.org/10.1007/978-3-662-62308-4_8

Keywords: Sentiment analysis · Framework · Predictive · Tourism · Cloud computing

1 Introduction

The tourism industry is an important sector of the world economy. In 2018, global tourism increased 6% according to the United Nations World Tourism Organization (UNWTO), and it is expected to in-crease by 3% to 4% in 2019. Despite the political and economic conflicts that continually arise around the world, global tourism metrics continues to rise. According to the Ministry of International Commerce and Tourism (MINCETUR in Spanish), in 2018, approximately 4.4 million foreign tourists visited Peru, 9.6% more than 2017, and generated US$ 4.9 billion of revenue [1].

SMEs constitute the majority of tourism service providers nationwide, however, they lack the infrastructure and resources to provide products and services to a greater number of people, since in order to be profitable they need to have a minimum of packages purchased. For this reason, if they do not manage to have this minimum of packages purchased, they need to outsource certain services to a larger tourism company and reduce their profit margins. In addition to reduced profits, there is considerable risk, as separating some of these services requires a non-refundable pre-deposit.

With this study, a framework is proposed that will predict the number of tourists who will visit an area and the packages they want to take, through a long-term sentiment analysis and a short-term, real-time analysis of Twitter posts. By applying sentiment analysis to these comments and posts, the system estimates a general level of satisfaction with the destinations visited and can thus estimate an increase or decrease in tourists during specific time periods. These statistics will allow SMEs to significantly reduce the risk of not reaching the minimum of packages purchased, and will even allow them to start building personalized tour packages as they will meet the interests of tourists, and can improve their profit margins.

2 Literature Review

2.1 Analysis of Different Languages

In the relevant research of language analysis, [3] developed a sentiment analysis system for the two most used languages in Malaysia, English and Malay, focusing on the lexicon, as the majority of the published research on sentiment analysis has concentrated on the vocabularies of the English lexicon. However, [4] presented a language-independent sentiment analysis model, with the domain based on n-grams of characters to improve classifier performance using the surrounding context. The results confirmed that this approach of integrating the surrounding context was more effective for data sets of different languages and domains. This suggests that a model based on n-grams of characters for data sets of multiple domains and languages is effective. Thus, a simple all-in-one classifier, that uses a mix of labeled data in multiple languages (or domains) to train a model of sentiment classification, can compete with more sophisticated domain or language adaptation techniques. On the other hand, [5] presents an innovative solution that considers space and temporal dimensions, using automatic geolocation techniques, for sentiment analysis of users that have a sense of belonging to a group. Geolocation is language independent and does not make previous assumptions about the users.

In the case of the articles mentioned above, the outcomes show that the sentiment analysis can be reliable. However, the proposed methods in [3,4] and [5] produce average accuracy due to the use of the slang, abbreviated words and dialects widely used in social networks and thus difficult to decipher.

2.2 Sentiment Analysis in the Tourism Industry

In relation to sentiment analysis in the tourism industry, [6] and [7] designed a model to analyze hotel customers' comments, [8] analyzed the flight experience of airline passengers from their social network comments, and [9] applied different sentiment analysis approaches for tourism in general, reviewing and evaluating them in terms of data sets used and performance relative to key evaluation metrics.

Nevertheless, most of the available hotel review or flight experience text data sets lack labels. As they represent feelings, attitudes and opinions that are commonly full of idiomatic expressions, onomatopoeias, homophones, phonemes, alliterations and acronyms, they are difficult to decipher and require a large amount of work to pre-process [6–8]. In particular, [8] uses sentiment analysis techniques to analyze negative, neutral and positive feelings in relation to the top ten airlines in the United States. And [9] outlines future research in tourism analysis as part of an expansive, Big Data approach.

2.3 Social Network

Regarding social networks, [10] developed a Sentiment Analysis Engine (SAE) that estimates the sentiment of users in terms of positive, negative or neutral polarity. Their SAE is based on the classification of an automatic text learning model, trained by real data flows deriving from different social network platforms that specialize in user opinion (for example, TripAdvisor). Monitoring and sentiment classification are then carried out on the continuously extracted comments from publicly available social networks such as Facebook, Twitter and Instagram, a procedure that [11] performs as well. In a specific case, [12] presents a model for analyzing the impact of a brand by fusing real data collected from Twitter over a 14-month period, and also analyzes the revisions that covers the existing methods and approaches in the sentiment analysis. In a more general case, [13] suggests a framework consisting of analysis modules and linguistic resources where two main analysis modules are run by a classification algorithm that automatically assigns class appropriate labels of intent and sentiment for a given text. However, [14] demonstrates that addressing negation can improve the final system and thus developed an unsupervised polarity classification system, based on the integration of external knowledge. To evaluate this influence, a group of tweets were first analyzed by their suggested unsupervised polarity classification system to detect negation, and then under a sentiment analysis that considered their detected negation, and a control that didn't. As seen above, traditional sentiment analysis emphasizes the classification of web comments in positive, neutral and negative categories. However, [15] goes beyond classification of sentiments by focusing on techniques that can detect the specific topics that correspond to positive and negative opinions. Combining these techniques can help understand the general reach of sentiment as well as sentiment drivers. Contrary to the articles previously mentioned, [16] analyzes the textual content as well as the visual one. As the old saying goes, "a picture is worth a thousand words", and the image tweet is a great example of a multimodal sentiment.

In conclusion, each article reviewed here has a different approach in analyzing social network sentiment, as they attack the problem from their individual perspective. For example, [16] focuses on the sentiment analysis based on visual and multimedia information. The results obtained in [14] reveal that the analysis of negation can greatly improve the accuracy of the final system. And [11] concludes that information extraction techniques based on Twitter allow for the collection of direct answers from a target public, and therefore provide a valuable understanding of public sentiment to predict an overall opinion of a specific product. In order to train its classification model, [13] suggests the linguistic resources of corpus and lexicon. Corpus consists of a collection of texts manually labeled with the appropriate classes of intention and sentiment. Lexicon consists of general terms of opinions and clusters of words that help to identify the intentionality and the sentiment. This later process requires manual entry of a large quantity of information and is therefore quite complicated and time-intensive.

2.4   Types of Neural Networks

In recent years, deep artificial neural networks, including recurrents, have won numerous pattern recognition and machine learning competitions. [17] summarizes succinctly the significant work of the last millennium. Shallow and deep learners are distinguished by the depth of their credit allocation routes, which are the chains of possibly learnable, causal links between actions and effects. Deep supervised learning, non-supervised learning, reinforcement learning, evolutive calculation, and indirect search of short programs that codify big and deep networks are reviewed. [18] looks to provide a complete tutorial and survey about recent developments, with the objective of enabling the efficient processing of deep neural networks (DNNs). Specifically, it provides a general vision of DNNs and analyzes several hardware platforms and architectures that can run them. It also summarizes the various development resources that allow researchers and professionals to get started in the field. [19] and [20] each developed Sentiment Analysis (SA) based on experiments in different Convolutional Neural Network (CNN) configurations, with [19] implemented on Hindi movie reviews collected from newspapers and online websites. The dataset was manually annotated by three native hindi speakers for model training preparation and experiments were carried out by using different numbers of convolution layers with a variable quantity and size of filters. [21] presents a similar model to [19], with neural convolution networks. However, the original model of convolution neural networks ignores sentence structure, a very important aspect of textual sentiment analysis. For this reason, [21] adds the association of parts to the convolution neural network, which allows the model to understand the sentence structure. To counteract a lack of data and actually improve the model's generalization capacity, [21] employs a generative adversarial network to obtain the common characteristics associated with emotions. Also, [22] suggests a sentiment classification model with a convolutional neural network that uses representations of several words to represent words that have not been pre-trained. The experimental outcomes of three data sets show that the suggested model, with an additional character-level integration method, improves the accuracy of the sentiment classification. On the other hand, [23] suggests a multiple attention network (MAN) for sentiment analysis that learns word and phrase level characteristics. MAN uses the vectorial representation of the input sequence as an objective in the first attention layer to locate the words that contribute to the sentiment of the sentence.

As has been shown, there are many models and configurations of neural networks, some more effective than others depending on their application and desired analysis. This is the case with [21], as opposed to [19], because it can overcome a lack of data availability. In the case of [23], where even though an individual word can indicate subjectivity, it can give insufficient context to determine the orientation of the sentiment. The authors posit that this sentiment analysis usually requires multiple steps of reasoning. Therefore, they applied a second attention layer to explore the information of the phrase around the key word.

3   Proposed Model

3.1   Model Analysis

To support SMEs, we present the following framework (Fig. 1) that analyzes information from social networks on a cloud platform to create a tourism preference metric that helps create desirable tourism packages. This is very important for SME tourist agencies because their business model relies on presenting desirable travel packages for travelers to visit tourist destinations. Historically, agencies used word of mouth information and in some cases, surveys, to design their travel packages, both of which lacked reliability. With the suggested framework, information is collected from social networks and review sites and processed in a low-cost, high performance, cloud platform that uses neural networks to analyze and calculate traveler sentiment. These outcomes are stored in a datastore where metrics/reports can be generated in the future. Performance indicators and representations of place and time trends will show where tourists feel more satisfied or motivated. With this information, tourist agencies can create better packages that reflect historic travelers' mood, and a more personalized sales environment.

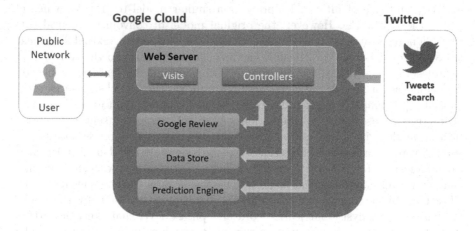

Fig. 1. Framework.

3.2   Components

Figure 2 shows the process of extraction of the information from Google and Twitter, first by a filter (hashtag), then the sentiment analysis to generate the charts that will present the trends and tourist satisfaction in relation to the different destinations.

Input. For data input we use two platforms: Twitter and Google Places/Maps. The Twitter platform will be used for measuring user motivation at a specific time, because this network fits with this type of analysis.

Twitter/Motivation The motivation data source will be Twitter, a platform commonly known for sharing people's mood or opinion at a specific moment and thus effective for measuring tourist motivation at different locations in real time. This information, related to trips, tourism, etc., will be collected in a massive way using related hashtags, which will provide the first source of data about tourist destinations during specific time periods.

Google Places/Maps Satisfaction. The second platform that will be used is Google Places/Maps. This network will be used for quantifying user satisfaction with registered places as this network focuses on opinions and levels of customer service. Google Places/Maps generates a more historical type of information as it doesn't operate with the real time aspect of Twitter, but provides us a more specific channel for information.

Cloud. To process the information, a Cloud platform will be used to help SMEs reduce operating costs (servers purchases, trainings, implementation, maintenance, etc.) and provide an market accepted processing speed and uptime standard of service.

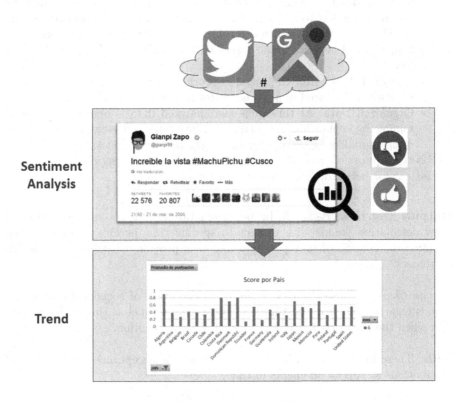

Fig. 2. Information analysis process.

Web Server.

– Tools.
  • Sent analysis. We will use neural networks for sentiment analysis on the massive information collected from Twitter and Google reviews to measure satisfaction/motivation for a specific moment and destination and ultimately quantify targeted users' moods. As the amount of data due to the number of opinions found in social networks is so large, the historical method of surveys or word of mouth opinions cannot possibly match the breadth of information from these networks. This information is vital for generating metrics and establishing trends.
  • Db. After being retrieved, the information will be stored in a database for persistence and accessibility for the application of our neural network

to assess our main parameters: travel trends and popular destinations, or those destinations that present an elevated positive emotion.

- • Neural network. For the sentiment analysis we use neural networks to calculate the emotion users display on social networks. The neural network for this study is Deep Feed Forward [24]. This network combines the wide and deep models to allow a high capacity of abstraction and processing speed. This network will use the previously extracted, transformed and loaded social network input data for its analysis.
- – System.
    - • Analyzed data. After the social network data is organized by the neural network, it is stored in a datastore for metric and trend analysis.
    - • Organized Data. At this stage the analyzed data is organized for data mining by separation/granularizating for time and location to create a decision-making spectrum for the creation of travel packages.
- – Safety Support. Safety support will manage user access/communication/ sessions for better organization and customer service.

Output. Organized data will be presented through charts and indicators to show destination trends with respect to specific time periods to improve the selection or creation of tour packages and thus provide a better experience to customers.

Trends Charts. These charts will show the evolution of social network users' moods regarding the tour destinations, presenting a timeline that will help predict when the travel packages could have the best reception.

KPIs. These indicators will be used to evaluate the acceptance across different destinations at specific moments in time.

4   Validation

4.1   Case

To validate the model presented, we used a case study to demonstrate that the proposal successfully solves the needs of SMEs in the Peruvian tourism market.

4.2   OT S.A.C. Company Information

OT S.A.C is a small tourist agency business that sells and distributes tour packages, with ten employees and monthly revenues of approximately $29,000. It is located in the district of Santiago de Surco, in the city of Lima, Peru. Its main suppliers are wholesale companies that provide it with a list of packages for sale and distribution. In turn, OT S.A.C. sells custom packages as requested by their clients. From its early stages, according to national regulations regarding tourist package distributors, this company benefited by its portfolio of existing clients.

As new companies entered the market in the same category and with the same products, price competition began and spurred a sudden growth in the sector.

Under these circumstances, the owners and managers focused on the use of new technologies to maintain or improve sales levels. That is why we proposed this emerging technology process model to OT S.A.C.

4.3   Implementation

In Fig. 3 the ETL (Extraction, Transform and Load) process is shown. This process is used for the ex-traction of data from the social networks, the transformation of those data, and the loading of those data into the database.

4.4   Program/APIs

For the framework implementation we developed a web tool that collects data from the Twitter and Google Places/Maps social networks using their respective APIs. With the Twitter API, tweets at a specific time can be collected and filtered by hashtags chosen for their relevance/closeness to tourism topics (Fig. 4). This data is filtered and cleaned of special characters, URLs, emoticons, and other factors that can negatively affect the sentiment analysis. Then, the data is sent to the cloud service where it will be processed and stored.

All reviews will collect the geographic location information they make reference to (Fig. 5). This information will be sent afterward to the API of the sentiment analysis tool. This will evaluate each request and will send the output of the analysis to the database API for future use.

4.5   Sentiment Analysis

For the implementation of sentiment analysis, 3 types of neural networks were compared: densely connected neural network (basic neural network), convolutional neural network (CNN) and short-term memory network (LSTM), which is a variant of the networks recurrent neural. To choose which type of neural network has the best performance, the 3 models were trained with the same database. The database used was [25], which contains sentences labeled with positive or negative feelings. In total there were 2748 sentences, which are labeled with a score, 1 (for positive) and 0 (for negative), as shown in (Fig. 6). The sentences come from three websites: imdb.com, amazon.com and yelp.com.

Densely Donnected Neural Network. The first neural network to be tested is a simple deep neural network. The embedment layer will have an input length of 100, the dimension of the output vector will also be 100, and the dense layer of 10,000 parameters. For the activation function, the sigmoid function was used. The Adam optimizer was used to compile the model. For the training of the neural network, 80% of the data was used, and 20% for validation. At the end of the training, the training precision was around 81.9% and the test precision

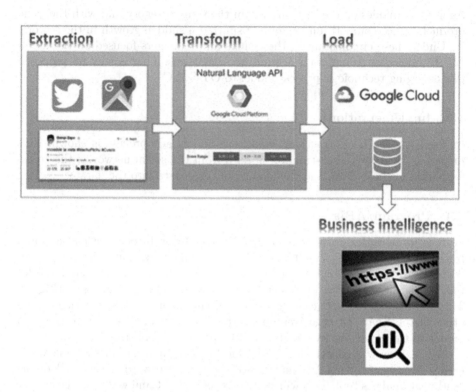

Fig. 3. ETL – Extraction, Transform, Load.

"#vacaciones", "#viaje", "#playas",
"#campo", "#diversion", "#travel",
"#summer", "#verano", "#invierno",
"#primavera", "#trekking", "#love",
"#hiking", "#tourism", "#tourist",
"#adventure", "#paisaje", "#trip", "#otoño",
"#holiday", "#journey", "#hotellif",
"#nature", "#sightseeing", "#gastronomia",
"#cruise", "#explore", "#travelblogger",
"#travel", "#travelphotography",
"#turismoecologico", "#outdoors"

Fig. 4. Hashtags used for the filtering.

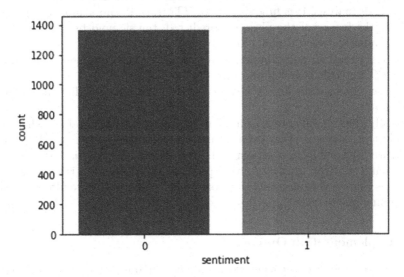

	Name/ID ↑	anio	ciudad	dia	locale	magnitud	mes	pais	puntuacion	tweet
☐	id=4785553677484032	2019	Mercedes	16	Mercedes, Argentina	0	6	Argentina	0	#SabadoDeFlojera #Lluvia #invierno #Casa...
☐	id=4787621469356032	2019	Armação dos	15	Armação dos Búzios	0.2000000029802	6	Brazil	0.100000001490	Yes, We're the couple in the airport shuttle wi...
☐	id=4793051281096704	2019	Cuauhtémoc	10	~	2.2000000476837	6	Mexico	0.699999988079	Beautiful morning, new friends, everything is ...
☐	id=4796213115224064	2019	Valparaíso	10	~	0.6999999880790	6	Chile	0.699999988079	Outfit #modaloretosaez #tendencias2019 # ...
☐	id=4796400818716672	2019	San Pedro Ch·	10	~	1.5	6	Mexico	0.100000001490	El magnifico Popocatépetl...................
☐	id=4803152373088256	2019	Chessy	14	Chessy, France	0.8000000119209	6	France	0.400000005960	Disneyland paris #travel #travelphotography...
☐	id=4805618086969344	2019	Caracas	10	~	0.8000000119209	6	Venezuela	0.400000005960	Habla por si sola #Lunes #Señales #AMF_C...
☐	id=4805841895030784	2019	Boston	16	Boston, MA	1.2999999523162	6	United State	0.300000011920	Girls just want to have fun! Hanging out to ce...
☐	id=4810099885342720	2019	La Pampa	16	La Pampa, Argentina	1.7000000476837	6	Argentina	0.5	My girl on a sunset in the countryside. @lapa...
☐	id=4814221644660736	2019	Guadalajara	10	~	1.2999999523162	6	Mexico	0.600000023841	Mi mamá siempre me decía que antes de sal...
☐	id=4814786122481664	2019	Almada	12	Almada, Portugal	0.8000000119209	6	Portugal	0.800000011920	Chilling ✎ #lisboa #yosoylafigura #juniorlafi...
☐	id=4815098690404352	2019	Almada	12	Almada, Portugal	0.8000000119209	6	Portugal	0.800000011920	Chilling ✎ #lisboa #yosoylafigura #juniorlafi...
☐	id=4821787967750144	2019	Tlahuiltepa	17	Tlahuiltepa, Hidalgo	0.8000000119209	6	Mexico	0.200000002980	Vi tantas Lunas... #vida #poquitaropa @DOB...
☐	id=4822.....	2019	Valparaiso	16		0.400000000790	6	Chile	0.400000000070	Outfit #lunes #modaloretosaez #look #m...

Fig. 5. Filtered, cleaned and stored tweets.

Fig. 6. Positive and negative sentences.

was 73.2% (Fig. 7). This means that the model is over-fitted, this occurs when the model performs better in the training set than the test set. Ideally, the performance difference between sets should be minimal.

Convolutional Neural Network (CNN). The convolutional neural network is a type of network that is mainly used for the classification of 2D data, such as images. A convolutional network tries to find specific characteristics in an image. Convolutional neural networks have also been found to work well with text data. Although text data is one-dimensional, 1D convolutional neural networks can be used to extract features from our data.

The created CNN has 1 convolutional layer and 1 grouping layer. The one-dimensional convolutional layer has 128 neurons. The kernel size is 5 and the activation function used is sigmoid. As can be seen in (Fig. 8), the training precision for CNN is 92.5%, and the test precision 83%.

Recurrent Neural Network (LSTM). Lastly, the LSTM, which is a network that works well with sequence data such as text, which is a sequence of words, will be tested. In this case, the LSTM layer will have 128 neurons, just like CNN. As can be seen in (Fig. 9), the training precision is 86% and the test precision is 85%, higher than that of CNN.

The result shows that the difference between the precision values for the training and the test sets is much smaller compared to the simple neural network and CNN. Furthermore, the difference between the loss values is also insignificant, therefore, it can be concluded that LSTM is the best algorithm for this case.

For text analysis, it is first cleaned of HTML code and symbols (Fig. 10)

Then the feeling of the phrase is predicted. the sigmoid function predicts a floating value between 0 and 1. If the value is less than 0.5, the sentiment is considered negative, and if it is greater than 0.5, the sentiment is considered positive. The result is as follows (Fig. 11):

The sentiment value for the instance is 0.87, which means that the sentiment is positive.

Finally, these results are presented in web-based statistical tables, where users can select the statistics they need for making their decisions (Fig. 12).

Figure 13 shows a chart generated by the system, where satisfaction is shown in different countries, where 1 is a very positive comment and -1 is a very negative one. This chart has been generated from 5,000 tweets that were filtered to show only comments made in June 2019.

4.6   Implementation Outcomes

As a result of the use of the framework, data collected from social networks is processed by the sentiment analysis tool and organized so that users can understand the tourism trends that occur during specific time periods. Through charts or a trend line, users will be able to see when the highest positive sentiment occurs for a specific destination, and thus present travel packages with a high probability of desirability and acceptance. Through this framework, the travel packages presented will have a higher probability of customer satisfaction. The outcomes of Twitter analysis we will be able to rate the users' mood or motivation across time. The outcomes from Google Reviews analysis will concern satisfaction levels, helping quantify the quality level of services offered at different destinations. These two metrics, generated through the experiences shared on social network, will help agencies design better travel packages.

4.7   Segmentation

The scenarios presented were generated from the data collected throughout the project. Our first scenario presents the picture of travel and tourism sentiment

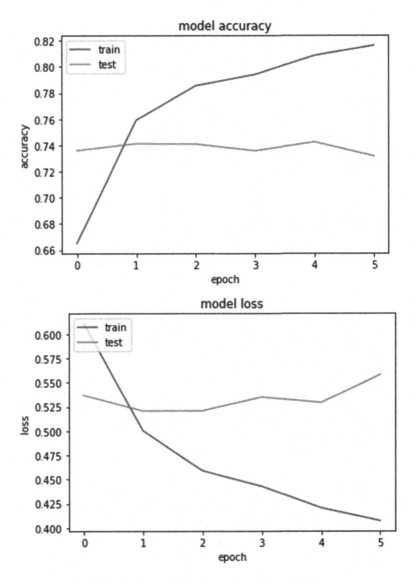

Fig. 7. Densely donnected neural network.

from different countries. The next scenario shows the number of tweets about specific tourism topics, which reveals to us the mood of a country regarding those topics. The last shows us raw data that companies can use to build their charts and metrics. With these scenarios, we can have a wider dataset from which to analyze tourism topics from different countries across time (Fig. 14).

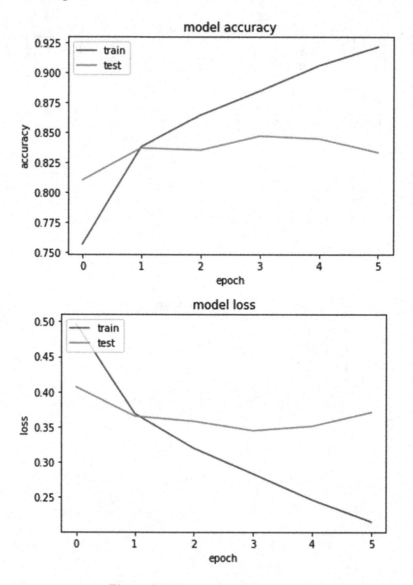

Fig. 8. Convolutional neural network

Scenario 1 – Average Ranking of Countries per Month. The preceding chart shows the average ranking obtained from tweets collected in June and processed through the sentiment analysis tool. It shows what countries have presented a positive attitude related to travel and tourism on social networks. From this, we can visualize which countries have a better reception to tourism topics and can evaluate which periods of time would be better for trip planning and travel package presentation (Fig. 15).

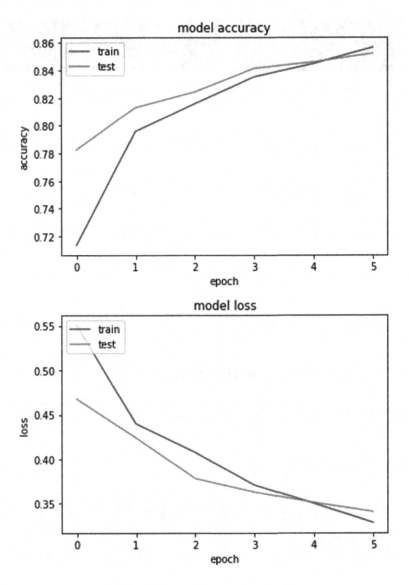

Fig. 9. Recurrent neural network

The best trip of my life.

 Mexico is the best country, with its stunning beaches, very friendly people and the food leaves you speechless

The best trip of my life Mexico is the best country with its stunning beaches very friendly people and the food leaves you speechless

Fig. 10. Clean text.

array([[0.87998194]], dtype=float32)

Fig. 11. Result of sentiment analysis.

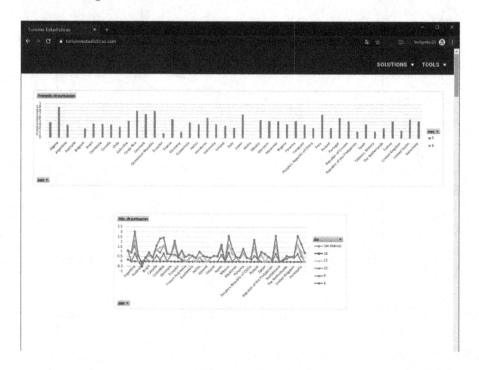

Fig. 12. Web platform.

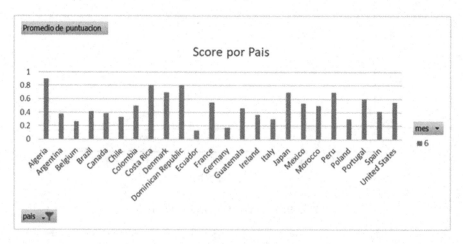

Fig. 13. Generated chart.

Scenario 2 – Sum of the Score of Countries per Day. In this chart, we show the sum of the sentiment analysis scores of tweets on specific days, by country. The chart is segmented by days, to present a more detailed vision of the sentiment in different countries on travel and tourism topics. The score is

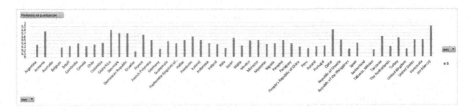

Fig. 14. Average ranking of countries per month.

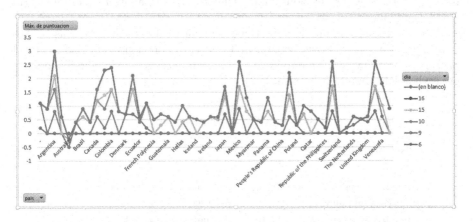

Fig. 15. Sum of the score of countries per day.

added to give a dimension to the number of tweets generated and to see what amount of people are discussing the topic in social networks. By taking into account the number of tweets generated and the scores obtained, we can have an idea of the tourism picture in different countries. With this, companies can make better decisions as to what are the best times to organize tourist packages (Fig. 16).

año	ciudad	dia	magnitud	mes	pais	puntuacion	tweet
2019	A Coruña	18	0.200000003	6	Spain	0.100000001	Winter is coming!! #verano #Coruña
2019	Abbotsford	10	0.5	6	Canada	0.200000003	The Guardian Raises the Level of #ClimateAlarmism to New Heights #IPCC #WMO #climatechange #globalwarming #science #gl
2019	Acapulco de	9	1	6	Mexico	-0.200000003	Calma . . . #travelstories #traveler #traveltheworld #laguna #lagoon #acapulco #guerrero #samsunga9 #vacacionas #vacation #nub
2019	Acapulco de	16	1.600000024	6	Mexico	0.5	Buenos días acapulquito!! . . . #travel #viajar #trip #viaje #viajes #instatravel #viajeros #turismo #travelphotography #viajando #beac
2019	Acapulco de	17	0.899999976	6	Mexico	0.899999976	Los "Nachitas" ☺☺☺☺ #Amigos #Familia #Compayes #Fiesta #Celebración #Viaje #Cumpleaños #Junio2019 #México #Acapulc
2019	Adelaide	10	0.899999976	6	Australia	0.200000003	Be curious..... about Nepal. #nepal #kathmandu #pokhara #trek #annapumacircuit #everestbasecamp #thedontforgettravelgroup #t
2019	Advanced D	10	2.799999952	6	United State	0.5	Are you on Summer Vacay yet? We want to hear about your #summer plans! We're celebrating Father's Day this Sunday, our 5-`
2019	Alcobendas	16	0.899999976	6	Spain	0.400000006	Si vives correctamente los Sueños vendrán a ti.□ #primavera #stronger #actitudpositiva #smile #smiling #Madrid #love #blogger #r
2019	Alegría	17	1.399999976	6	Republic of	0.300000012	Thank God for these creation. 🍴 Thank God for the experience. 🍵 Thank God we all survived. 🍽 #Summer #Vacay #Cebu #Cany

Fig. 16. Analyzed Tweets.

Scenario 3 – Analyzed Tweets. This chart shows an abstract from the obtained tweets. The table shows the time where the tweet was collected and its sentiment analysis. The Magnitude is the numeric equivalent of the intensity of the sentiment shown in the tweet, while the Score shows how positive or negative the tweet has been rated. With this information we can generate charts and metrics to help with company decision-making.

4.8   Case Study Result

After the process model implementation and thanks to the support of the web tool, OT SAC reduced decision-making time by 60%, to an average of 6 days (in a 3–8 day range).

Due to the support provided by the sentiment analysis, OT SAC was able to reduce logistic expenditures, as seen in Table 1. In addition to increasing travel package sales, they were able to create their own packages, reducing dependence on a wholesale company (See Table 2).

Table 1. Loss by logistic expenditures.

Loss caused by logistical expenses ($)			
Before		After	
Annual expenses caused by price difference	4824.24	Annual expenses caused by price difference	3600
Annual loss caused by purchases	11,104.3	Annual loss caused by purchases	7773.01
Loss %	57%	Loss %	27%
Indicator	Critic	**Indicator**	Positive

Table 2. Before and after sales comparison.

Sales comparison ($)			
Before		After	
Separation of Package with Wholesaler	10%	Separation of Package with Wholesaler	0%
Coordination with transport	5%	Coordination with transport	5%
Average sales	6	Average sales	13
Total price	$ 2810	Total price	$ 2810
Gain per package (20%–10%)	$ 281	Gain per package (20%)	$ 562
Total gain	**$1686**	**Total gain**	**$ 7384**

Regarding the results, we can say the following:

Before implementing the process model, the company lost customers for not responding to requirements on time, and lost package reservations due to not considering clients' needs. However, after implementation, not only could customers' needs be more effectively considered, but the company was able to negotiate more favorable rates with suppliers that will generate a greater profit margin.

5 Discussion

The neural network has been trained with sentences in English, however, it can be trained with any other language. An already analyzed database was used, since it requires a large number of already analyzed sentences to train the neural network, however, for it to perform better it is recommended to train it with more updated comments and using slangs, so that it can analyze the comments with more precision.

The proposed system does not interpret emojis, and in many comments the sentiment information is contained in these, that is why, it is recommended to assign a word to each emoji and train the network with these values so that you can later interpret the emojis.

6 Conclusions

The results show that the framework will help SME tourist agencies use historical data and sentiment analysis to offer more desirable, customized travel packages.

The generation of charts and trends will vary according to the user necessity, by place, time, etc., and is immediate since all of the information required is organized and stored in a datastore in the cloud.

Some tweets do not present any sentiment by text, but rather through a picture attached to the tweet. Further research should investigate adding a system of sentiment analysis of images to the developed text-based framework to provide an even more accurate analysis.

References

1. Confederación Nacional de Instituciones Empresariales Privadas - CONFIEP (2019). Turismo en Perú (21 Noviembre del 2019). Recuperado de https://www.confiep.org.pe/noticias/economia/turismo-en-peru/
2. Zapata, G., Murga, J., Raymundo, C., Dominguez, F., Moguerza, J.M., Alvarez, J.M.: Business information architecture for successful project implementation based on sentiment analysis in the tourist sector. J. Intell. Inf. Syst. 53(3), 563–585 (2019). https://doi.org/10.1007/s10844-019-00564-x
3. Zabha, N., Ayop, Z., Anawar, S., Erman, H., Zainal, Z.: Developing Cross-lingual Sentiment Analysis of Malay Twitter Data Using Lexicon-based Approach. Int. J. Adv. Comput. Sci. Appl. (2019). https://doi.org/10.14569/IJACSA.2019.0100146
4. Kincl, T., Novák, M., Pribil, J.: Improving sentiment analysis performance on morphologically rich languages: language and domain independent approach. Comput. Speech Lang. 56, 36–51 (2019). https://doi.org/10.1016/j.csl.2019.01.001
5. Fernández-Gavilanes, M., Juncal-Martínez, J., Méndez, S., Costa-Montenegro, E., Castaño, F.: Differentiating users by language and location estimation in sentiment analisys of informal text during major public events. Expert Syst. Appl. 117 (2018). https://doi.org/10.1016/j.eswa.2018.09.007
6. Zvarevashe, K., Olugbara, O.: A framework for sentiment analysis with opinion mining of hotel reviews, pp. 1–4 (2018). https://doi.org/10.1109/ICTAS.2018.8368746

7. Gunasekar, S., Sudhakar, S.: Does hotel attributes impact customer satisfaction: a sentiment analysis of online reviews. J. Glob. Scholars Mark. Sci. **29**, 180–195 (2019). https://doi.org/10.1080/21639159.2019.1577155

8. Anitsal, M.M., Anitsal, I., Anitsal, S.: Is your business sustainable? A sentiment analysis of air passengers of top 10 US-based airlines. J. Glob. Scholars Mark. Sci. **29**, 25–41 (2019). https://doi.org/10.1080/21639159.2018.1552532

9. Alaei, A., Becken, S., Stantic, B.: Sentiment analysis in tourism: capitalizing on big data. J. Travel Res. **58**, 004728751774775 (2017). https://doi.org/10.1177/0047287517747753

10. Ducange, P., Fazzolari, M., Petrocchi, M., Vecchio, M.: An effective Decision Support System for social media listening based on cross-source sentiment analysis models. Eng. Appl. Artif. Intell. **78**, 71–85 (2019). https://doi.org/10.1016/j.engappai.2018.10.014

11. Suresh, H., Gladston, S.: An innovative and efficient method for Twitter sentiment analysis. Int. J. Data Min. Model. Manage. **11**, 1 (2019). https://doi.org/10.1504/IJDMMM.2019.096543

12. Vural, A., Cambazoglu, B., Karagoz, P.: Sentiment-focused web crawling. ACM Trans. Web **8**, 2020–2024 (2012). https://doi.org/10.1145/2396761.2398564

13. Haruechaiyasak, C., Kongthon, A., Palingoon, P., Trakultaweekoon, K.: S-sense: a sentiment analysis framework for social media sensing. In: IJCNLP 2013 Workshop on Natural Language Processing for Social Media (SocialNLP), pp. 6–13 (2013)

14. Zafra, S.M., Martín-Valdivia, M., Martínez-Cámara, E., López, L.: Studying the scope of negation for spanish sentiment analysis on Twitter. IEEE Trans. Affect. Comput. **PP**, 1 (2017). https://doi.org/10.1109/TAFFC.2017.2693968

15. Cai, K., Spangler, W., Chen, Y., Li, Z.: Leveraging sentiment analysis for topic detection. Web Intell. Agent Syst. **8**, 291–302 (2010). https://doi.org/10.3233/WIA-2010-0192

16. You, Q.: Sentiment and emotion analysis for social multimedia: methodologies and applications, pp. 1445–1449 (2016). https://doi.org/10.1145/2964284.2971475

17. Schmidhuber, J.: Deep learning in neural networks: an overview. Neural Netw. **61** (2014). https://doi.org/10.1016/j.neunet.2014.09.003

18. Sze, V., Chen, Y.-H., Yang, T.-J., Joel, E.: Efficient processing of deep neural networks: a tutorial and survey. Proc. IEEE **105** (2017). https://doi.org/10.1109/JPROC.2017.2761740

19. Rani, S., Kumar, P.: deep learning based sentiment analysis using convolution neural network. Arab. J. Sci. Eng. **44**(4), 3305–3314 (2018). https://doi.org/10.1007/s13369-018-3500-z

20. Alharbi, A.S.M., de Doncker, E.: Twitter sentiment analysis with a deep neural network: an enhanced approach using user behavioral information. Cogn. Syst. Res. (2018). https://doi.org/10.1016/j.cogsys.2018.10.001

21. Du, C., Lei, H.: Sentiment analysis method based on piecewise convolutional neural network and generative adversarial network. Int. J. Comput. Commun. Control **4**, 7–20 (2019). https://doi.org/10.15837/ijccc.2019.1.3374

22. Ju, H., Yu, H.: Sentiment classification with convolutional neural network using multiple word representations 1–7 (2018). https://doi.org/10.1145/3164541.3164610

23. Du, T., Huang, Y., Wu, X., Chang, H.: Multi-attention network for sentiment analysis. In: NLPIR 2018: Proceedings of the 2nd International Conference on Natural Language Processing and Information Retrieval, pp. 49–54 (2018). https://doi.org/10.1145/3278293.3278295

24. Zapata, G., Murga, J., Raymundo, C., Alvarez, J., Dominguez, F.: Predictive model based on sentiment analysis for Peruvian SMEs in the sustainable tourist sector. In: IC3K 2017 - Proceedings of the 9th International Joint Conference on Knowledge Discovery, Knowledge Engineering and Knowledge Management, vol. 3, pp. 232–240 (2017)
25. Kotzias, D., et al.: From group to individual labels using deep features. In: KDD (2015)

Author Index

Printed in the United States
By Bookmasters